SHORT QUESTION & ANSWER SERIES

BOTANY

For BSc Pass and Hons students and for the students appearing for all major Indian competitive examinations

SHORT QUESTION & ANSWER SERIES
BOTANY

Dibyendu Saha MSc (Botany-Cytogenetics)

New Central Book Agency (P) Ltd
LONDON
HYDERABAD ERNAKULAM BHUBANESWAR
DELHI KOLKATA PUNE GUWAHATI

NCBA

REGD OFFICE
8/1 Chintamoni Das Lane, Kolkata 700 009, India
email: ncbapvtltd@eth.net

OVERSEAS
NCBA (UK) Ltd, 149 Park Avenue North, Northampton, NN3 2HY, UK
email: ncbauk@yahoo.com

EXPORT
NCBA Exports Pvt. Ltd, 212 Shahpur Jat, New Delhi 110 049
email: ncbaexp@ncbapvtltd.com

BRANCHES

212 Shahpur Jat
New Delhi 110 049
email: ncbadel@ncbapvtltd.com

Shop Nos. 3 & 4, Vinayak Towers
681/B Budhwar Peth, Appa Balwant Chowk
Pune 411 002
email: ncbapune@ncbapvtltd.com

Shop No. 15, Bharati Towers, Ground Floor
Forest Park, **Bhubaneswar** 751 001
email: ncbabub@ncbapvtltd.com

House No. 3-1-315, 1st Floor
Nimboliadda, Kachiguda, **Hyderabad** 500 027
email: ncbahydb@ncbapvtltd.com

GSS Shopping Complex, 1st Floor
Opposite College Post Office, Convent Road
Ernakulam 682 035
email: ncbaernk@ncbapvtltd.com

Radhamayee Complex (Second Floor)
Opp. SBI, Panbazar, Jaswanta Road
Guwahati 781 001
email: ncbaguwahati@ncbapvtltd.com

SHORT QUESTION & ANSWER SERIES: BOTANY • Dibyendu Saha

First Published: January 2007
Reprinted: 2009, 2013

PUBLISHER AND TYPESETTER
New Central Book Agency (P) Ltd
8/1 Chintamoni Das Lane, Kolkata 700 009

PRINTER
New Central Book Agency (P) Ltd
Web-Offset Division, Dhulagarh, Sankrail, Howrah

COVER DESIGNER
Soumen Paul

COVER PRINTER
Biswanath Studio, Kolkata

PROJECT TEAM
Prabhat Jas, Subas Maitra

ISBN 978 81 7381 528 7

Price: ₹ 125.00

Dedicated to my
parents
and
beloved students

CONTENTS

Preface

It is of immense pleasure to me to author the book SHORT QUESTION & ANSWER SERIES: BOTANY based on the undergraduate (Hons) and postgraduate level syllabi of different major universities across the country. It is indeed one of the most intricate tasks to pen a comprehensive book on Botany; and more so, when it has to be restricted to a limited syllabus. It would, at the same time, be preposterous to assume that such a vast arena of the subject could be condensed within a limited number of pages. Judged from this perspective, this book is only a humble endeavour to help the young aspirants, desirous of getting through an examination that has SHORT ANSWER TYPE QUESTIONS as one of its important components.

However, while compiling the book, utmost care has been taken and sincere efforts have been made to make it the best guide of its kind. Nevertheless, we cannot claim that the book has attained perfection; nor do we make any claim to originality—the ideas expressed, information incorporated and principles expounded are those of renowned scholars and researchers in the discipline.

Any suggestion for improvement will be cordially acknowledged and respectfully incorporated in future editions. We will deem our endeavour amply rewarded if the book proves to be beneficial to those for whom it is intended.

Last but not least, I owe a lot to Mr Amitabha Sen, the Publisher, but for whose inspirational support it would not have been possible for me to give a concrete shape to what I had envisaged. It has been a pleasure for me to work with Mr Prabir Ghosh, who took great care to produce the book.

Finally, I request the readers to send their suggestions, criticisms, and comments for further betterment of the book, and we can assure everyone that they will be thankfully received and appreciated.

10 October 2006 **DIBYENDU SAHA**

ACKNOWLEDGEMENTS

At the very outset I wish to express my heartfelt gratitude and indebtedness to the following:

Sri Chandra Sekhar Mukhopadhyay, Secretary, Rathindranath Mazumdar, Headmaster, Asim Goswami, Assistant teacher and all the other staffs of Santinagar Vidyamandir (Higher Secondary); Dr Mihir Das, Ex-Prof. Serampore College, Hooghly; Prof. Dr Prosanto Kumar Bhattacharya, HOD of Botany, The University of Burdwan, Burdwan; Dr Kalyan Kr Dey, HOD of Botany, Darjeeling Govt College, The University of North Bengal; Dr Nimai Chandra Saha, Principal, Jhargram Raj College, The University of Vidyasagar; Dr Rabindranath Bhattacharya, Presidency College, The University of Calcutta; Sri Tapas Mallick, Teacher, St Paul's Mission School, Kolkata; Achyutananda Bhandari, E.C.L. Employee, Burdwan; Bipasa Mondal, MSc student, The University of Kalyani; Bonani Banerjee, MSc student, Darjeeling Govt College, The University of North Bengal; Tinku Das (Ex), Supravat Maji, Atanu Pal, students, B.B. College, Burdwan; Kaushik Banerjee, Biswajit Mondal, student, T.D.B. College, Burdwan and all other former and present students.

I am extremely thankful to Subhra Chattaraj, MSc student, The University of Vidyasagar for valuable cooperation at the time of manuscript preparation.

I am grateful to Sri Amitabha Sen, Director, Sri S. Sinha Ray, CEO, Sri Soumen Paul and Sri Prabir Ghosh, New Central Book Agency and Sri Subas Maitra, Editor for making this book a reality.

1 INTRODUCTION

Botany is the branch of biology which deals with the study of plants of the present and the past in all their aspects. The term "BOTANY" is derived from the Greek work *botane* which means herb in plants.

The Three Kingdoms : Three PROTISTA classification established by Haeckel (1866). Based on the selection and evolutionary theory of Charles Darwin (1859).

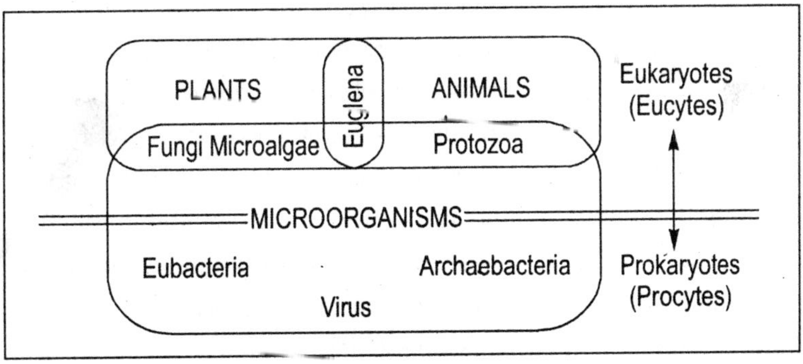

The Three Kingdoms : Plants, animals and microorganisms. The differoces between eukaryotes and prokaryotes

Plant kingdom (Protista) : At present about **3,71,745** species of different plants, angiospermic plants **2,86,000** species, gymnosperms **640** species, pleridophytes **10,000** species, bryophytes **23,000** species, fungi **40,400** species and algae **8,675**, have been discovered and identified by botanists.

The plant kindom is fundamentally divided into –

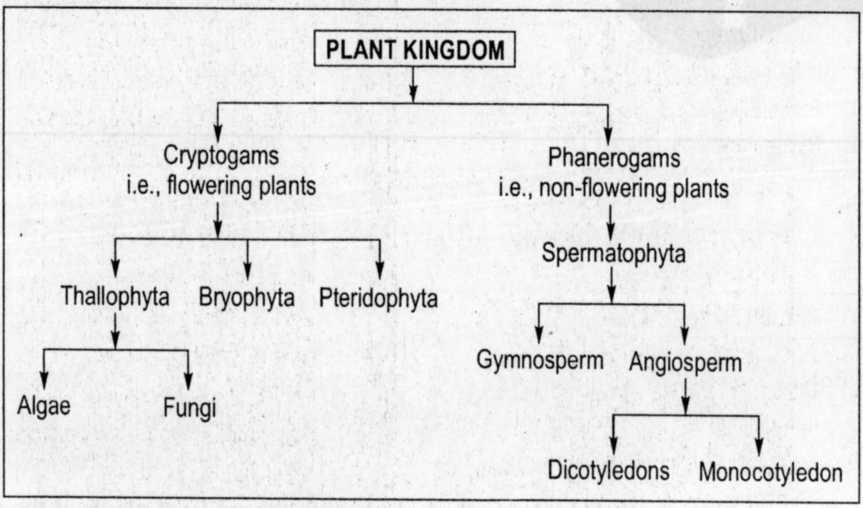

GENERAL INFORMATION

MAJOR SUBDIVISIONS OF BIOLOGY

	Name	Study of
1.	Morphology	form and body structure of animals
2.	Anatomy	internal structures and systems as visible to naked eye
3.	Histology	tissues and cells under microscope
4.	Cytology	cell structure, its organelles & their functions
5.	Physiology	functions of various organ systems of the organism
6.	Biochemistry	biological actions, in relation to chemical principles
7.	Endocrinology	endocrine glands, their secretions (hormones) and their actions
8.	Embryology	(Developmental biology) — Study of development, growth & repair of organisms
9.	Genetics	inheritance of variations in the organisms
10.	Evolution	origin of life and gradual differentiation or descent of species
11.	Ecology	relationship between organism & its environment
12.	Palaeontology	fossils, their occurrence & distribution in time
13.	Palaeozoology	fossil animals
14.	Taxonomy	classification of animals and plants, their evolutionary relationship with other organisms
15.	Zoogeography	distribution patterns of animals on the earth in time & space
16.	Limnobiology	fresh water ponds, lakes & streams

	Name	Study of
17.	Zoopathology	various disease of animals
18.	Immunology	resistance of organisms to infections
19.	Radiobiology	effects of radiation of life
20.	Psychology	the behaviour of an organism in its environment

IMPORTANT BRANCHES OF BIOLOGY

	Branch	Study of
(A)	**Botany**	
1.	Aerobiology	Organisms & their parts present in the air
2.	Actinobiology or Radiobiology	Effect of radiation on organisms
3.	Agronomy	Management of farms, cultivation of field crops
4.	Agrostology	Grasses
5.	Algology or Phycology	Algae
6.	Anatomy or Internal morphology	Internal structure
7.	Anthology	Flowers
8.	Arboriculture	Cultivation of ornamental trees & shrubs
9.	Autecology	Ecology of an individual species & its population (Including the effect of other organisms & environmental conditions on every stage of its life cycle)
10.	Bacteriology	Bacteria
11.	Biochemistry or Phytochemistry	Chemicals constituting the body & chemical reactions going on therein
12.	Biometrics or Biostatistics	Application of statistics for analysing biological problems
13.	Bionics	Technical applications of mechanisms devised by living nature
14.	Biophysics	Physical aspects of living systems
15.	Biotechnology	Utilization of living organisms or of substances obtained from them in industrial processes
16.	Bryology	Bryophytes
17.	Carcinology	Malignant tissues, tumors, cancers
18.	Cell Biology	Structure, function, reproduction & life history of cells
19.	Chemotaxonomy	Interrelations & classification of plants on the basis of chemicals present in their body
20.	Chronobiology	Functions of organisms in relation to time
21.	Cryobiology	Life at low temperatures
22.	Cytology	Detailed structure of cell
23.	Cytogenetics	Cytological basis of inheritance
24.	Dendrology	Shrubs and trees

	Branch	Study of
25.	Dendrochronology	Estimation of age of woody plants
26.	Ecology	Reciprocal relationship between organisms & their environment
27.	Economic Botany	Usefulness of plants & plant products
28.	Edapho(bio)logy or Pedo(bio)logy	Life in soils
29.	Embryology	Events leading to the fertilisation of egg and development of embryo
30.	Enzymology	Enzymes
31.	Epidemiology	Occurrence & distribution of diseases
32.	Ethnobotany	Relationship between primitive people & plants
33.	Ethology	Typical species (often instinctive) behaviour of organisms
34.	Etiology	Causes of diseases
35.	Evolutionary biology	Origin of life & descent of advanced forms of organisms from the primitive ones
36.	Exobiology or Xenobiology	Living organisms in space or at planets other than earth
37.	Floriculture	Cultivation of flower yielding plants
38.	Forestry	Development, utilisation & conservation of forests & forest products
39.	Genecology	Genetic diversity within a species in relation to climate & soil
40.	Genetics	Heredity & variations
41.	Genetic Engineering	Genetic manipulations to produce an organism with a new combination of properties
42.	Gnotobiotics	Germ-free life
43.	Histochemistry	Chemical nature of tissues
44.	Histology	Detailed structure of tissues
45.	Horticulture	Propagation of plants for fruits & ornamental value
46.	Immunology	Natural or acquired resistance to specific diseases
47.	Karyology	Nucleus
48.	Lichenology	Lichens
49.	Limno(bio)logy	Life in fresh water
50.	Microbiology	Miroorganisms
51.	Molecular biology	Molecular aspects of life processes (Nature of molecules in the cell & their interaction)
52.	Molecular genetics	Molecular basis of mode of inheritance
53.	Monerology	Monera
54.	Morphology	Form & structure
55.	Morphogenesis	Changes occuring during development and their differentiations
56.	Mycology or Mycetology	Fungi
57.	Numerical taxonomy	Utilization of principles of mathematics for determining taxonomic relationship
58.	Olericulture	Cultivation of vegetables

	Branch	Study of
59.	Oncology	Tumors
60.	Ontogeny	Embryonic development stages
61.	Palaeobotany	Plant life as it existed in the past (based on study of fossils i.e. remains of ancient life)
62.	Parsitology	Parasites
63.	Pathology or Phytopathology	Diseases – their symptoms, causes, spread and control, etc.
64.	Polynology	Pollen grains
65.	Pharmacognosy	Uses of plants & their parts in drug industry
66.	Pharmacology	Preparation of drugs & their effects on organisms
67.	Phenology	Changes in plant communities with seasons
68.	Photobiology	Effects of light
69.	Phylogeny	Evolutionary relationship among different groups of organisms
70.	Physiology	Processes & functions associated with life
71.	Phytogeography	Distribution of plants on the surface of earth
72.	Plant breeding	Production or creation of improved varieties by selection, mating and other procedures
73.	Pomology	Fruits & fruits yielding plants
74.	Protistology	Protists
75.	Pteridology or Pterology	Pteridophytes
76.	Serology	Serum & antigen antibody interactions
77.	Silviculture	Cultivation of timber yielding plants
78.	Space biology	Life in the space
79.	Synecology	Ecology of plant communities
80.	Spermology or Seed biology	Seeds
81.	Systematic Botany or Systematics	Diversity & kinds of organisms & evolutionary relationships among these (includes taxonomy and phylogeny)
82.	Taphonomy	Conditions conduction to the fossilisation of organisms in the past
83.	Taxonomy	Classification, arrangement, identification & nomenclature (naming)
84.	Teleology	Explanation of phenomena in terms of need of organisms (e.g., stem bends towards light because it needs light) not scientifically accepted
85.	Teratology	Abnormal embryonic development
86.	Toxicology	Production of toxins & their effects on organisms
87.	Trophology	Nutrition
88.	Virology	Viruses
89.	Xylotomy	Xylem vessels
90.	Zymology	Fermentation

SOME INTERESTING PLANTS

1.	Tallest plant (Tree)	*Eucalyptus amygdalina* (Australian)
2.	Smallest plant	*Lemna*
3.	Biggest plant (Tree)	*Sequoiodendron giganteum.*
4.	Oldest tree	*Macrozamia*
5.	Smallest flower	*Wolfia microscopica*
6.	Smallest seed	*Orchid*
7.	Smallest fruit	*Wolfia*
8.	Largest fruit	*Lodoicea malderica*
9.	Largest leaf	*Victoria amazonica*
10.	Smallest leaf	*Wolfia*
11.	Largest flower	*Rafflesia arnoldi*
12.	Largest flower in India	*Sapira*
13.	Largest pitcher plant	*Darlingtonia*
14.	Smallest cell like structure	*Mycoplasma gallisepticum* or *Dialister pneumosintes* (bacteria)
15.	Biggest ovule	*Cycas cercinalis*
16.	Biggest antherozoid	*Cycas*
17.	Smallest gymnosperm	*Zamia pygmaea*
18.	Smallest pteridophyte	*Azolla*
19.	Largest chromosomes	*Trillum* (30μ)
20.	Plant with longest leaves	*Raphia*
21.	Maximum number of chromosomes	*Opioglossum* (fern) (2n = 1260)
22.	Minimum number of chromosomes	*Haplopappus gracilis* (2n = 4)
23.	Smallest parasite	*Arceuthaobium*
24.	Hardest wood	*Hardwichia binata*
25.	Lightest wood	*Ochroma lagopus balsa*
26.	Shootless plant	*Rhizophora* or *Semicarpus*
27.	Biggest cactus	*Cereus giganteus*
28.	Longest sea-weed	*Macrocystis pyrifero*
29.	Biggest orchid	*Gramatophyllum speciosum*
30.	Rootless angiosperm	*Aldsovanda* or *Taeniophyllum*
31.	Heaviest wood	*Olea laurifolia*
32.	Biggest bacterium	*Spirillum*

SOME IMPORTANT ABBREVIATIONS

A	Androecium (Stamen)
Å	Angstrom unit (0.001 of a micron)
ABA	Abscisic Acid
ADP	Adenosine Di-phosphate
ADH	Antidiuretic Hormone (Vasopressin)
AAAA	Autotetraploid
ACTH	Adinocorticotropin Hormone
AIDS	Acquired Immune Deficiency Syndrome
AMP	Adinosine Monophosphate
ATP	Adinosine Triphosphate
BCG	Bacillus Calamette-Gurein Vaccine (For T.B.)
BMR	Basic Metabolic Rate
BOD	Biochemical Oxygen Demand
BNHS	Bombay Natural History Society
CO-A	Coenzyme A
COD	Chemical Oxygen Demand
CAT	Computer Assisted Tomography
CNS	Central Nervous System
CSF	Cerebro Spinal Fluid
CITES	Conservation on International Trade in Endangered Species (of wild life Fauna & Flora)
CPCB	Central Pollution Control Board
DNA	Deoxyribonucleic Acid
DPD	Diffusion Pressure Deficit
DPN	Diphospho-pyridine Nucleotide
2-4-D	2-4 dichiorophenoxy Acetic Acid
Ebr	Ebracteate
ER	Endoplasmic Reticulum
ECG	Electrocardiogram
EMG	Electromyogram
ETS	Electron Transport System
F^+	Bacterium with fertility factor
F_1	First filial generation
FAD	Flavin Adenine Dinucleotide
FAN	Flavin Adenine Nucleotide
FMN	Flavin Adenine Mononucleotide
FSH	Follicle Stimulating Hormone
GA	Gibberellic Acid
GAP	Glyceraldehyde Phosphate
GTP	Guanosine Triphosphate
GFR	Glomerular Filtration Rate
GH	Growth Hormone—Somatotropin
GRH	Growth-releasing hormone
GTH	Gonadotropic Hormone
HFr	High Frequency of gene recombination
HIV	Human Immunodeficiency Virus
HMP	Hexose Mono Phosphate
HLA	Human Leukocyte Antigen
IAA	Indole-3 Acetic Acid
IBA	Indole-3 Butyric Acid
IMR	Infant Mortality Rate
IUD	Intra-Uterine Device, placed in uterus prevents implantation of fertilized ovum
IBP	International Biological Programme
IBWL	Indian Board for Wild Life
IUCN	International Union for Constervation of Nature and natural resources
LH	Leuteinising Hormone–Leutotropin
LRH	Leuteinising Hormone–Releasing Hormone
LSD	Lysergic Acid Dimethylamide
MAB	Man and Biosphere Programme
MBC	Maximum Breathing Capacity
NAA	Naphthalene Acetic Acid
NAD	Nicotinamide Adenine Dinucleotide
NADH	Reduced NAD
NADP	Nicotinamide Adenine Dinucleotide Phosphate
NADPH	Reduced NADP
NGF	Nerve Growth Factor
NMNH	National Museum of Natural History
NWAP	National Wild Life Action Plan
nm	Nanometer (10^{-9} meter)
OGF	Ovarian Growth Factor

OP	Osmotic Pressure
PAN	Peroxyacetyl Nitrate
PEP	Phosphoenol Pyruvic Acid
PGAL	Phosphoglyceraldehyde
PGA	Phosphoglyceric Acid
PPLO	Pleuroneumonio-like organism
PRH	Prolactin Releasing Hormone
P/R ratio	Photosynthesis/Respiration ratio
RAD	Radiation Absorbed Dose
RDS	Respiratory Distress Syndrome
REF	Renal Erythropoeitic Factor
Rh	Factor – Rhesus group of red cell agglutinogens
RPF	Renal Plasma Flow
RNA	Ribonucleic acid
R.Q.	Respiratory Quotient
RUBP	Ribulose Biphosphate
rRNA	Ribosomal RNA
SP	Suction pressure
SPCA	Proconuertin – clotting factor VII
STH	Somatotropin growth hormone
TCA	Tricarboxylic acid Cycle or Kreb's Cycle
TMV	Tobacco Mosaic Virus
TP	Turgor Pressure
TPN	Triphospho Pyridine Nucleotide
tm	Melting point of DNA
tRNA	Transfer RNA
U	Uracil
URF	Uterine Relaxing Factor–Relaxin
UVR	Ultra Violet Rays
VD	Venereal disease
UNDP	United Nations Development Programme
WP	Wall Pressure
WPSI	Wildlife Preservation Society of India
WFN	Wildlife Fund for Name
WWF	World Wildlife Fund

FATHER OF DIFFERENT BRANCHES OF BIOLOGY

1.	Father of Biology	Aristotle
2.	Father of Botany	Theophrastus
3.	Father of Modern Botany	Linnaeus
4.	Father of Bacteriology	Leeuwenhock
5.	Father of Immunology	Edward Jenner
6.	Father of Genetics	G. J. Mendel
7.	Father of Modern Genetics	T. H. Morgan
8.	Father of Microbiology	Louis Pasteur
9.	Father of Cytology	Robert Hooke
10.	Father of Mycology	Micheli
11.	Father of Plant Physiology	Stephen Hales
12.	Father of Indian Palaeobotany	Prof. Birbal Sahni
13.	Father of indian Bryology	Prof. S. R. Kashyap
14.	Father of Indian Mycology	Prof. E. J. Butler
15.	Father of Indian Plant Embryology	Prof. P. Maheshwari
16.	Father of Indian Phycology	Prof. M. O. P. Iyenger

17.	Father of Indian Plant Genetic	Prof. B. P. Pal
18.	Father of Indian Ecology	Prof. R. Mishra
19.	Father of Special Creation Theory	Father Sanrez
20.	Father of Zoology	Aristotle
21.	Father of Evolutionary Ideas	Empedocles
22.	Father of Eugenics	Francis Galton
23.	Father of Mutation	Hugo de Vries
24.	Father of Modern Embryology	Karl Ernst Von Baer
25.	Father of Palaeontology	Leonardo da Vinci
26.	Father of Taxonomy	Carolus Linnaeus
27.	Father of Blood Groups	K. Landsteiner
28.	Father of Blood Circulation	William Harvey
29.	Father of Comparative Anatomy	G. Cuvier
30.	Father of Modern Genetics	Bateson
31.	Father of Medicine	Hippocrates
32.	Father of Radiation Genetics	H. J. Muller
33.	Father of Microscopy	Marcello Malpighi
34.	Father of Endocrinology	Thomas Addison

NOBEL PRIZE WINNERS IN MEDICINE AND PHYSIOLOGY 1990-2006

2005 *Barry J. Marshall* and *J. Robin Warren* – discovery of the bacterium *Helicobacter pylori* and its role in gastritis and peptic ulcer disease.

2004 *Buck Linda Axel Richard* – multigene family may encode odorant receptors.

2003 *Paul Lauterbur* and *Peter Mansfield* – to develop magnetic resonance imaging (MRI).

2002 *Sydney Brenner, H. Robert Horvitz* and *John E. Sulston* – gene regulation of organ development and progammed cell death or apoptosis.

2001 *Leland H. Hartwell, R. Timothy Hunt* and *Paul M. Nurse* – regulators of the cell cycle.

2000 *Arvid Carlsson, Paul Greengard* and *Eric Kandel* – signal transduction in the nervous system.

1999 *Gunter Blobel* – protein have intrinsic signals that govern their transport and localization in the cell.

1998 *Robert Francis Furchgott, Louis J. Ignarro* and *Ferid Murad* – nitric oxide as a signalling molecule in the cardiovascular system.

1997 *Stanley Ben Prusiner* – Prions – a new biological principle of infection.

1996 *Peter C. Doherty* and *Rolf M. Zinkernagel* – specificity of the cell-mediated immune defence.

1995 *Edward B. Lewis, Christiane Nusslein-Volhard* and *Eric F. Wieschaus* – genetic control of early embryonic development.

1994 *Alfred Goodman Gilman* and *Martin Rodbell* – G protein and the role of these proteins in signal transduction in cells.

1993 *Richard John Roberts* and *Phillip A. Sharp* – split gene.

1992 *Edmond H. Fischer* and *Edwin G. Krebs* – reversible protein phosphorylation as a biological regulatory mechanism.

1991 *Erwin Neher* and *Bert Sakmann* – function of single ion channels in cells.

1990 *Joseph E. Murray* and *E. Donnall Thomas* – organ & cell transplantation in the treatment of human disease.

2006, 2 AMERICANS WIN NOBEL PRIZE IN MEDICINE AND PHYSIOLOGY

Americans **Andrew Z. Fire** and **Craig C. Mello** won the Nobel Prize in Physiology and Medicine, discovering a powerful way to turn off the effect of specific genes, opening a new avenue for disease treatment.

"RNA interference" is already being widely used in basic science as a method to study the function of genes and it is being studied as a treatment for infections such as the AIDS and hepatitis viruses and for other conditions, including heart disease and cancer.

2 MICROBIOLOGY

Q. 1. What is prophage?

Ans. In lysogenic cycle the viral genome becomes integrated with the bacterial genome – in this integrated state the viral genome is called prophage.

Q. 2. What is flagella?

Ans. Thread-like locomotor appendages extending outward from the plasma membrane and the cellwall are called flagella. Most motile bacteria move by use of the flagella.

Q. 3. What is monostrichous bacteria?

Ans. When bacteria have only one flagellum located at the end it is said to be monostrichous. E.g. *Pseudomonas aeruginosa.*

Q. 4. What is amphitrichous bacteria?

Ans. When bacteria have a single flagella located at each pole it is called amphitrichous. E.g. *Aquspirillium serpens.*

Q. 5. What do you mean by flagellin?

Ans. The filament of flagella of bacteria is a hollow, rigid cylinder constructed to a single protein called flagellin (Mw 30,000 to 60,000)

Q. 6. What is pili?

Ans. Pili are tiny, hallow projections. They are used to attach bacteria to sufrace and are not involved in movement.

Q. 7. What is pilin?

Ans. A pili is composed of subunit of protein called pilin.

Q. 8. What is conjugated pili or sex pili?

Ans. Conjugated pili or sex pili are found only in certain groups of bacteria, attach two cell and may furnish a pathway for the transfer of genetic material DNA in the process of conjugation.

Q. 9. What is fimbriae?

Ans. Fimbriae are type of pili. They help bacteria adhere to surface – such as cell-surface – and the interface of water and air.

Q. 10. What do you mean by pellicle?

Ans. Some aerobic bacteria form a shiny or fuzzy thin layer at the air-water interface of a growth culture. This layer is called a pellicle.

Q. 11. Name a bacteria used in alcoholic fermentation.

Ans. *Zymnomonas mobilis.*

Q. 12. Name two bacteria used in vinegar production.

Ans. *Acetobactor, Gluconobactor.*

Q. 13. What is biofertiliser?

Ans. Biofertiliser define as biologically active products or microbial inoculants or bacteria, algae and fungi which may help biological nitrogen fixation, solubilisation of phosphorus and potash and also decomposing organic wastes for the benefit of plants.

Q. 14. Name two symbiotic bacteria.

Ans. *Rhizobium, Azospirillum.*

Q. 15. Name two free-living bacteria.

Ans. *Azotobactor, Azospirillum.*

Q. 16. Name two endospore forming bacteria.

Ans. *Bacillus, Clostridium* sp.

Q. 17. What is virion?

Ans. It is a unit of virus or the complete mature virus particle.

Q. 18. What kind of nucleic acid is generally found in plant virus?

Ans. RNA i.e. ribonucleic acid found in plant virus.

Q. 19. What is peptidoglycan?

Ans. This is the chemical compound present in the cell wall of bacteria. It is an enoromous polymer composed of identical subunits. The polymer contains two sugar derivatives – N-acetyle glucosamine (NAG) and N-acetylemuramic acid (NAM).

Q. 20. What is endospore?

Ans. A number of gram-positive bacteria can form a special resistant dormant structure called endospore. Endospore develops within vegetative bacterial cell of several genera – *Bacillus, Clostridium* sp.

Q. 21. What is nucleoid?

Ans. The most striking difference between prokaryot and eukaryot is that the prokaryotic chromosome is located in an irregular shaped region called nucleoid.

Q. 22. Write down two living features of virus.

Ans. (i) Multiplication of virus particles takes place only by replication rather than by growth and division.

(ii) Viruses contain either DNA or RNA but never both.

Q. 23. What is mesosomes?

Ans. The infolding of the bacterial cell membrane are known as mesosome. The mesosome performing the function of DNA replication and septum formation in bacterial cell.

Q. 24. What are bacterial ribosome?

Ans. Ribosomes are small particles, 10-30μ in diameter. It is 70s type, lie free in cytoplasm composed of RNA and protein.

Q. 25. Name two photosynthetic bacteria.

Ans. *Chlorobium* sp., *Chlorobacterium* sp.

Q. 26. What is plasmid?

Ans. Inside the bacterial cell are present some extra genetic material which are circular in nature and called plasmid. The plasmid mainly consists of non-functional DNA sequence.

Q. 27. What are the differences between virion and viroid?

Ans.

Virion	*Viroid*
1. A complete virus particle.	1. A transmissible agent that cause a number of plant or animal diseases.
2. It has capsid.	2. It has no capsid.
3. It has two types of nucleic acid — DNA and RNA.	3. It has only one type of nucleic acid — RNA.

Q. 28. Why bacterial genome is called genophore?

Ans. Bacterial cells are prokaryotic in nature. Hence they do not possess a well-defined nucleus – so the genetic material remains diffused throughout the cytoplasm of the cell – hence bacterial genome is called genophore.

Q. 29. What is capsomere?

Ans. The individual unit of protein subunit of capsid is known as capsomere.

Q. 30. Name two RNA virus.

Ans. *Polio* virus, *Reo* virus.

Q. 31. What are the T-even bacteriophages?

Ans. T_2, T_4 and T_6 bacteriophages are known as T-even bacteriophages.

Q. 32. What are the basic differences between a bacterial chromosome and a plant chromosome?

Ans.

Bacterial chromosome	*Plant chromosome*
1. This chromosome is not capable of showing mitotic or meiotic division.	1. It shows mitotic as well as meiotic division.
2. No formation of chromatid.	2. Here chromatid formation occurs.

Q. 33. Name two DNA virus.

Ans. Herpes virus, chicken-pox virus.

Q. 34. Name two bacteria useful for agriculture.

Ans. *Azotobacter* sp., *Clostridium* sp.

Q. 35. What are pleomorphic bacteria?

Ans. Although most bacterial species have cells that are of a fairly constant characteristics shape, some have cells that can exibit a variety of shapes known as pleomorphic bacteria.

Q. 36. What is bacterial chromosome?

Ans. Bacterial cell does not contain characteristic nucleus as in eukaryotic cell. There is no evidence of nuclear membrane since the DNA of bacterial cell does not form a discrete nucleus and is not capable of meiotic and mitotic division.

Q. 37. Name two main components of bacterial cell wall.

Ans. Mucopeptide and polysaccharide.

Q. 38. What do you mean by temperate phage?

Ans. The type of growth cycle are seen in bacteriophages e.g. lytic and lysogenic. On the basis of such growth cycle, phage can be divided in two — temperate and virulent. Those phages exibit lysogenic cycle called temperate phage.

Q. 39. Write down the function of bacterial pili.

Ans. (i) It help in adhesion of bacterial cell to other cells.

(ii) This also helps entry of genetic material during mating.

Q. 40. What is bacterial capsule?

Ans. Some bacterial cells are surrounded by a viscous substance forming a covering layer or envelope around the cell wall known as bacterial capsule.

Q. 41. What is the function of capsule?

Ans. (i) It may provide protection against temporary drying by binding water molecules.

(ii) It may block attachment of bacteriophage.

(iii) They may be antiphagocytic.

Q. 42. Write down the commercial use of bacteria.

Ans. (i) Lactic acid production by *Lactobacillus* is used as food.

(ii) Vinegar production by *Mycoderma coli.*

(iii) *Clostridium butricum* is used for fiber extraction.

Q. 43. What is periplasmic space?

Ans. A space is seen between the plasma membrane and the outer membrane in electron micrographs of gram-negetive bacteria known as periplasmic space.

Q. 44. What are the basic differences between gram positive and gram negative cell wall?

Ans.

Gram-positive	*Gram-negative*
1. Gram-positive cell wall contain a single 20 to 80 nm thick homogeneous peptidoglycan layer	1. Gram-negative cell wall contains 2 to 7 nm peptidoglycan layer.
2. Periplasmic space absent	2. In between plasma membrane and outer membrane there presents periplasmic space
3. O site chain or O antigen absent.	3. O site chain or O antigen present.

Q. 45. What is phylogenetic classification?

Ans. The classification which is bassed on 16s-rRNA sequence known as phylogenetic classification. E.g. Bacterial classification of Atlas.

Q. 46. Name two gram-positive bacteria.

Ans. *Bacillus* sp., *Streptomyces* sp.

Q. 47. Mention two bacteria which are used as biofertiliser.

Ans. *Azotobacter* sp., *Rhizobium* sp.

Q. 48. What are thermophytes?

Ans. The strong class Archaeobacteria consist of some bacteria which are tolerant of acid; these are known as thermophytes.

Q. 49. Define biopesticides.

Ans. Biopesticides are define as biological organism like bacteria, virus, fungi and plant product which can be used to kill a suceptible insect of plant. They are called biopestiside.

Q. 50. How does plasmid differ from episome?

Ans. Plasmid is the extra genetic material in bacterial cell. It is typically short, usually circular, double-stranded.

The bacterial plasmid that have ability to integrate into host chromosome and are called episome.

Q. 51. Name one free-living N_2-fixing bacteria.

Ans. *Azotobacter* sp.

Q. 52. How bacterial ribosome differ from eukaryotic ribosome?

Ans. In bacterial ribosome it is 70s type made up of 50s and 30s — two subunits.

But in eukaryotic organism there are 80s ribosome made up of 60s and 40s subunits.

Q. 53. Give an example of pleomorphic bacteria.

Ans. *Mycoplasma* sp.

Q. 54. What is P.H.B.?

Ans. P.H.B is one of the most common inclusion body consisting of poly β hydroxybutyric acid, contains β-hydroxybutyrate molecules joined by ester bonds between the carboxyl and hydroxyl groups of adjacent molecules.

Q. 55. What is capsid?

Ans. The protein coat of a virus is known as capsid.

Q. 56. Some informations.

1. Smallest bacteria	*Mycoplasma pneumonea*
2. Largest bacteria	*Thiomargarita namibiensis*
3. Gram-positive bacteria	*Bacillus* sp.
4. Gram negative bacteria	*Salmonella toyphimurium*
5. Photosynthetic bacteria	*Chlorobium* sp.
6. N_2-fixing bacteria	*Azotobactor* sp.
7. Symbiotic bacteria	*Rhizobia* sp.
8. Free-living bacteria	*Azolla* sp.
9. Pleomorphic bacteria	*Mycoplasma* sp.
10. Endospore forming bacteria	*Clostridium* sp.
11. Rod-shaped bacteria	*Bacillus* sp.
12. Filamentous bacteria	*Streptomyces* sp.
13. Stalked bacteria	*Galionella*
14. Monotrichous bacteria	*Pseudomonas aeruginosa.*
15. Amphitrichous bacteria	*Aquaspirillum serpens.*
16. Lopostrichous bacteria	*Pseudomonas fluorescens.*
17. Peritrichous bacteria	*Salmonella typhi*

Q. 57. Write down the chemical structure of NAM and NAG.

Ans.

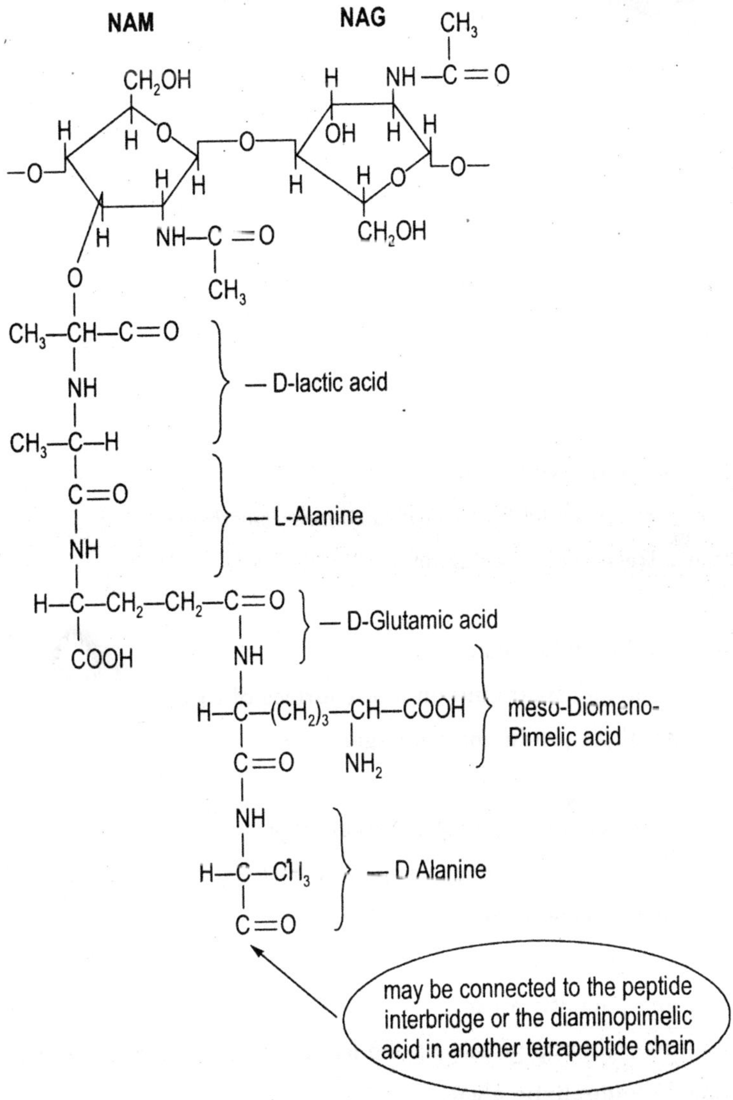

3 ALGAE

Q. 1. Name one epiphytic algae.

Ans. *Ptilota plumosa* which grows on *Laminaria hyperborea.*

Q. 2. Name a fresh water red algae.

Ans. *Compsopogon* sp.

Q. 3. Name a lithophytic algae.

Ans. *Batrachospermum* sp. (found on moist surface of rock).

Q. 4. Name an algae producing agar-agar.

Ans. *Gelidium, Gracilaria* etc.

Q. 5. Name four genera of terrestrial green algae.

Ans. *Vaucheria terrestris, Pleurococcus* sp., *Iwanoffia* sp., *Pritschiella* sp.

Q. 6. Name two marine algae.

Ans. *Sargassum* sp, *Laminaria* sp.

Q. 7. Name one parasitic algae.

Ans. *Cephaleuros* (chlorophyceae) on the leaves of *Magnolia* and *Camellia sinensis.*

Q. 8. Name an endophytic algae.

Ans. *Nostoc* sp. in thallus of *Anthoceros* sp.

Q. 9. Name some edible algae.

Ans. *Chlorella* sp. *Ulva lactuca.*

Q. 10. From where is "Carrageenin" produced?

Ans. Carrageenin is produced from *Chondrus crispus* (red algae).

Q. 11. Name some iodine-producing algae.

Ans. *Laminaria* sp., *Fucus* sp.

Q. 12. Name some nitrogen-fixing algae.

Ans. *Oscillatoria princeps, Anabaena fertilissima.*

Q. 13. Which algae have role in medicine?

Ans. *Laminaria japonica, Gelidium* sp.

Q. 14. From where is `Vermifuge' obained?

Ans. Vermifuge is obtained from *Digehea, Codium* sp.

Q. 15. What is algin? What is its economic importance.

Ans. The extract of alginic acid in the form of sodium alginate is usually called algin. Algin is used in making icecream, and also in the manufacture of adhesive and artificial silks.

Q. 16. What is chantransia stage?

Ans. In some members of Rhodophyta the carpospore is liberated from the carposporangium and germinates into a heterotrichous filament which morphologically is quite different from the main thallus. This heterotrichous filament is known as chantransia stage. E.g. – *Batrachospermum* sp.

Q. 17. What do you mean by `red snow'?

Ans. In some algae there is present some asexual reproductive structure called hypnospore. The wall of hypnospore sometime become red due to the presence of haematochrome and represent a phenomenon know as `red snow' when it mixed with the snow of Arctic, Alpine rigion. E.g. – *Sphaerella.*

Q. 18. What is akinate?

Ans. The vegetative cells of certain filamentous algae become elongated and thick-walled. These are known as akinate. The akinate contains abundant reserve food. E.g. *Cladophora* sp. *Pithophora* sp.

Q. 19. What is palmella stage?

Ans. Under adverse condition in some algae such as *Chlamydomonas*, the contents of the vegetative cell divides to form new daughter cell. This increases in number by repeated cell division. All the cells remain clustered together inside the mucilage envelope formed by the gelatinisation of the parent cell walls of the successive generations. This is known as palmella stage.

Q. 20. What is coenobium? Give an example.

Ans. In some Cyonophyceae and Chlorophycean members there form colony. The colony is of constant size and shape. A colony consists of a definite number of cells arranged in a specific manner and forming an integrated whole structure called coenobium. E.g. *Volvox* sp., *Gonium* sp.

Q. 21. What are the chemical components of cell walls of Cyanophyceae?

Ans. The cell walls of Cyanophycean members consist of various substances like muramic acid, glucosamine, alanine, glutamic acid and diaminopimilic acid.

Q. 22. What type of pigment is present in Cyanophyceae?

Ans. The pigment phycobilin is present in the class Cyanophyceae. The pigment is divided into two parts –

(i) c-phycocyanin, and (ii) c-phycoerythrin.

Q. 23. What do you mean by blepharoplast?

Ans. In some algae there is a single granule at the base of each flagellum of a motile cell. This is known as blepharoplast.

Q. 24. What is eye-spot?

Ans. Certain motile, unicellular green algae possess a structure which under microscope appears as a small bright reddish or brownish red spot. It is known as the `eye-spot'

Q. 25. What is nucleoplasm?

Ans. In Myxophyceae DNA microfibrils are irregularly present in cytoplasm – that is called nucleoplasm which is somewhat transparent having lower electron opacity. The nucleoplasm only contains DNA fibrils. Histone and protamins are absent.

Q. 26. What do you mean by isokont and heterokont?

Ans. When a motile cell of algae provides equal flagella then it is called isokont and the unequal flagella is called heteroknot.

Q. 27. Name some green algae which produce aplanospore.

Ans. *Ulothrix* sp, *Microspora* sp.

Q. 28. Name some algae of water pollution.

Ans. *Ulothrix zonata, Amphora veneta* etc.

Q. 29. Name one algae possessing 2n chromosome number in its vegetative state.

Ans. *Fucus* sp and *Sargassum* sp.

Q. 30. Difference between trichome and filament.

Ans. Trichome is multicellular vegetative body not covered by any sheath. When the trichomes are covered by the sheath then they are known as filament.

Q. 31. Distinguish between acronematic and pantanematic flagella.

Ans. The acronematic is whiplash flagellum and has a smooth surface while the tinsel, i.e. pantonematic flagellum, bears longitudenal rows of fine minute hairs arranged along the long axis almost to the tip of the flagellum.

Q. 32. What is nannandrium?

Ans. In certain species of *Oedogonium,* anthridia develop on a few celled filament and consists of a basal stalk cell and 2-3 antheridial cells, this filament is called – nannandrium.

Q. 33. What are the stored food of Phaeophyceae?

Ans. The stored food of phaeophyceae is laminarin and mennitol.

Q. 34. Name one algae with isomorphic alteration of generation.

Ans. *Ectocarpus arabicus.*

Q. 35. Name a prokaryotic algae.

Ans. *Anabaena circinalis.*

Q. 36. Why is *Vaucheria* placed in Xanthophyceae?

Ans. (i) Absence of chlorophyll-b and presence of zoospore.

(ii) Presence of two kinds of flagella – one is whiplash and the other is tensil type on the sperm.

(iii) Presence of chlorophyll-e.

Q. 37. What is fucoidin?

Ans. On Phaeophycean cell the outer cell wall contains certain colloidal substances due to presence of which the cell wall become gelatinous. They are known as fucoidin.

Q. 38. What is diatomin?

Ans. In the member of Bacillariophyceae there is present a special golden colour pigment in the diatom cell called diatomin.

Q. 39. What is Cyanophycean granule?

Ans. In Cyanophycean cell the structural granules which are sorrounded by thylakoids are large dense granules which usually occur near the cross-wall and contain phospholipids known as cyanophycean granule.

Q. 40. What is gonimoblast filament?

Ans. In *Polysiphonia,* from the auxillary cell there develops a small lateral outgrowth called gonimoblast initial. This process is repeated and ultimately a set of gonimoblast initial are formed. Each initial develops into a short multicellular filament known as gonimoblast filament.

Q. 41. Which algae is called superfood of the future?

Ans. *Chlorella ellipsoidea* is called superfood of the future.

Q. 42. In which algae tetraspore is found?

Ans. *Polysiphonia platycarpa.*

Q. 43. Name the organism which produces carpospore.

Ans. *Polysiphonia* sp.

Q. 44. What is gridle lamella?

Ans. The cell wall of the diatoms consists of two overlapping halves or valves. The silicified portion, i.e. the edges of the two valves, are adjoined and held in place by each forming a side-piece or connecting band known as gridle lamella.

Q. 45. What is androspore?

Ans. In the nanandrous species of *Oedogonium* the dwarf male filament bearing antheridia are produced by the germination of antherozoid-like zoospore called androspore.

Q. 46. Do the members of Cyanophyceae reproduce sexually?

Ans. The true sexuality does not exist in the member of Cyanophyceae but a kind of parasexual phenomenon known as genetic recombination. This first appeared in *Anacystis nidulans* by Kumar in 1992 is found.

Q. 47. What is heterotrichous habit? Give an example from algae.

Ans. In some members of the algae the plant body is differenciated into a prostrate system of branched filamentous body and erect system of branched filamentous body that growing away from the substratum. This is called heterotrichous habit. E.g. *Batrachospermum* sp.

Q. 48. Distinguish between raphe and pseudoraphe.

Ans. The raphe is furrow extending length-wise through the wall of valve of diatom cell. But in some members of order pennales (*Fragilaria*) there is no true raphe, but the cell may possess a clear longitudenal area in the median region of the valve known as pseudoraphe.

Q. 49. Distinguish between pennate and centric diatoms.

Ans. (i) The Centric diatoms are radially symmetrical and Pinnate diatoms are bilaterally symmetrical.

(ii) In Centrales, auxospore are formed mainly by oogamy and in Pennals auxospore formed mainly by isogamy.

(iii) Statospore or microspore may be formed in Centrales but in Pennales statospore or microspore are never formed.

Q. 50. What is kelp? States its economic importance.

Ans. The seaweeds belonging to brown algae are commonly known as kelp. They contain soda, potash, iodinic acid. The soda and potash obtained from kelp are used in the manufacture of soap, glasswere and alum. High-yielding ioidine are used in chemotheraphy of thyroid and in the treatment of goiter.

Q. 51. What is triphasic alternation of generation?

Ans. The life cycle of *Polysiphonia* possesses three phases – a haploid gametophyte, a diploid carposporophyte and a free-living diploid tetrasporophyte. That is why *Polysiphonia* exibit a distinct alteration of generation known as triphasic alternation of generation.

Q. 52. What is heterocyst? Name one genus bearing heterocyst.

Ans. Certain vegetative cells of some filamentous algae become larger in size than the neighbouring cells. The walls of these type of cells somewhat thickend being composed of two layers – the outer of pective substance and the inner of cellulose called heterocysts. E.g. – *Nostoc commune.*

Q. 53. Name an algae causing plant disease. Mention its family.

Ans. *Cephaleuros virescens* is the cause of 'red rust of tea' of tea plant (*Thea sinensis*). The family is Trentepohliaceae.

Q. 54. What is floridean starch?

Ans. In the member of Rodophyceae the reserve food occurs in the form of small grains of a peculiar solid carbohydrate called floridean starch. It is a glycogen-like polysaccharide which turns yellow or brown with iodine.

Q. 55. What is conceptacle?

Ans. In some members of Phaeophyta, the swollen tips of the thalli the receptacles, which lack midrib are covered with small scattered pimple-like projections with small openings which lead into cavities, known as conceptacle. E.g. – *Fucus* sp.

Q. 56. Mention the pigments found in Xanthophyta.

Ans. Chlorophyll-a, β-carotene, xanthophyll, lutein, neoxanthin.

Q. 57. What is an apical cap? Where do you find it?

Ans. In some members of Chlorophyceae the lateral wall of some cell of each filament have one or more ring-like marking at the distal end, those are called cap cell or apical cap. It is found in *Oedogonium* sp.

Q. 58. Comment on the economic importance of Cyanophyceae.

Ans. Blue green algae of the family Nostocaceae are capable of fixing atmospheric nitrogen – and increase the soil fertility. Some members of Cyanophyceae serve as food for fish, they are eaten by fish and by other aquatic organisms which, in turn, may be eaten by man. In this way the blue green algae constitute an important link of food chain of an ecosystem.

Q. 59. What are the pigments found in blue green algae?

Ans. In blue green algae are found chlorophyll-a, β-carotene, myxoxanthin, myxoxanthophyll, zeaxanthin, allophycocyanin, phycocyanin and phycoerythrin etc.

Q. 60. Describe the chromatic adaptation in algae.

Ans. The Cyanophycean members have power to light screens against intense illumination. The same species, when grown in different wave-length of light, show significant variation in pigment composition. This is called complimentary chromatic adaptation also called Gaidukov phenomenon.

Q. 61. What are the functions of heterocyst?

Ans. The functions of heterocyst are :

(i) It plays role in sporulation.

(ii) It is the site of nitrogen fixation, considered ecologically and agriculturally important in tropical rice field.

Q. 62. Name the pigments present in the members of Phaeophyceae.

Ans. Chlorophyll a and chlorophyll c, α and β carotens, xanthoplylls, diatoxanthin, flavoxanthin, fucoxanthin, lutein, vialoxanthin.

Q. 63. What are carpogonia? In which genus are they found?

Ans. The female sex organ of *Polysiphonia* are called the carpogonia which are produced terminally upon a greatly reduced fertile trichoblast of a female gametophyte and consists of a swollen base called carpogonial base. They are found in *Polysiphonia* sp.

Q. 64. Name two unicellullar green algae.

Ans. *Chlamydomonas* sp., *Chlorella* sp.

Q. 65. Name one halophytic algae.

Ans. *Dunalicla, Chlamydomonas chrenberghii.*

Q. 66. Name a algae which forms lichen.

Ans. *Nostoc, Trentepohlia* forms symbiotic association with fungi.

Q. 67. Which algae produce antibiotics?

Ans. Chlorellin from *Chlorella.*

Q. 68. What do you mean by water blooms?

Ans. In the rainy and spring season blue green algae, some green and golden brown algae become so abundant that the water in the ponds, lakes and reservoirs becomes cloudy and assumes a yellowish or greenish tinge. Sometimes a floating yellowish green scum may develop on the surface of water. Popularly we call these manifestations of algal growth water blooms. Algae that are involved are — *Microcystis, Aphanizomenon* etc.

Q. 69. Name one algal fossil.

Ans. *Aphanocapsa* sp (Cyanophyceae).

Q. 70. Distinguish between algae and fungi :

Ans.

Algae	*Fungi*
1. Mainly aquatic	1. Mainly terrestrial
2. R.F.M. — Starch	2. R.F.M. — Glycogen
3. Cell wall cellulosic	3. Cell wall chitaneous
4. Autotrophs	4. Heterotrophs

Q. 71. Which algal class is nutritionally known as obligate photoautotrops?

Ans. Cyanophyceae.

Q. 72. Where do you find pseudovacuoles? What is it?

Ans. The pseudo-vacuoles were discovered in certain planktonic blue green algae— *Anabaena* sp.

Healthy cells of many planktonic blue green algae contain numerous small bodies of irregular shapes known as pseudovacuales or gas vacuales.

Pseudovacuoles consist of packed array of minute membranous cylindrical gas-containing structure with closed conical ends. It is imperable to water but freely permeable to gases (oxygen, nitrogen, argon). In cyanophyceae, vacuoles are absent in the protoplast.

Q. 73. Name a harmful algae.

Ans. *Gymnodinium brevis.*

Q. 74. In which algae do you find a giant nucleus?

Ans. *Bryopsis hypnoides.*

Q. 75. What is neutral spore?

Ans. In some algae the protoplast of vegetative cells directly function as spore – these are neutral spores. E.g. *Asterocystis* sp.

Q. 76. Name a yellow green algae which produces aplanospore.

Ans. *Vaucheria* sp.

Q. 77. Name the sex organ of *Chara*.

Ans. In *chara* the female sex organ is – Globule

Male sex organ is – Nucule.

Q. 78. What type of ribosome is found in the cell of Cyanophyceae?

Ans. In cyanophyceae there is found 70s type of ribosome. It is made up of larger subunits of 50s and smaller subunits of 30s.

Q. 79. What are the trumpet hyphae?

Ans. In *Laminaria*, the cross-section of the strip shows three distinct regions, namely meristoderm, cortex, medullary region. Sometimes in medullary region certain cells are elongated and become considerably swollen at the septa. These cells are called trumpet hyphae or trumpet cell.

Q. 80. What is raphe?

Ans. In the members of Bacillariophyceae the cell wall consist of two halves. On the obverse side of each half, there are linear perforations forming longitudinal slits called raphe.

Q. 81. Compare unilocular and plurilocular sporangia.

Ans. The unilocular sporangia are oval-shaped and develops from the terminal cell of short lateral branches.

The plurilocular sporangia are elongated and develops from the terminal cell of lateral branches. E.g. *Ectocarpus* sp.

Q. 82. In which algae do you find tubers?

Ans. *Chara* sp., *Cladophora* sp.

4 FUNGI

Q. 1. What is mycorrhiza?

Ans. Sometime fungi also live in close association with the root of gymnosperms and angiosperms — this association between fungal hyphae and the root of higher plant is know as mycorrhiza.

Q. 2. What is the store product of fungi?

Ans. The store product of fungi is — Glycogen.

Q. 3. Name a fungi used as a food by human.

Ans. *Agaricus bisporus.*

Q. 4. Name two poisonous fungi.

Ans. *Amantia verna, Amanita virosa.*

Q. 5. Name two medicinal fungi.

Ans. *Penicillium notatum, P. chrysogenum.*

Q. 6. What is appressoria?

Ans. Some parasitic fungi produce flattened hyphal pressing organs from which minute infection pegs usually grow that penatrate the epidermal cell of the host. This organ is known as appressoria.

Q. 7. What is mycoparasite?

Ans. Many groups of heterotrophic organism involve a nutritional relationship that is favourable to the parasite, and usually unfavourable to host. The parasitism of one fungus on another is known as mycoparasitism and the parasitising fungus is the mycoparasite.

Q. 8. What is diplodisation?

Ans. A process by which a haploid cell or a mycelium or similar structure becomes diploid having conjugate nuclei is designated as diplodization.

Q. 9. What is sporophore?

Ans. Fructification of fungi is called sporophore. In Ascomycetes, the sporophore are called Ascocarp while in Basidiomycetes they are called basidiocarp.

Q.10. What is budding? Where do you find it?

Ans. In some fungi the vegetative i.e., parent cell, give rise small outgrowths called buds. Each bud separates from the parent cell and grow into a new individual. E.g. – yeasts.

Q. 11. Name one unicellular fungus.

Ans. *Saccharomyces cerevisiae.*

Q. 12. Name two industrial products produced by *Penicillium*.

Ans. *Penicillium roqueforti, Penicillium camemberti.*

Q. 13. What is rhizomorph?

Ans. Sometime the hyphae of some fungi fuse to each other forming a rope-like structures running parallel to each other. These thick, gelatinous, dark brown and rope-like coiled structures are called rhizomorphs.

Q. 14. What is heterocious fungi? Give an example of it.

Ans. When a fungus produces its different spores – forms on two distinct and distantly related hosts to complete its life cycle, it is called heterocious fungi. E.g. – *Puccinia graminis* requires wheat as one primary host and barberry plants as the altemate secondary host.

Q. 15. What do you mean by holocorpic and eucarpic fungi?

Ans. When the total plant body of fungi take part in reproduction those fungi are known as holocorpic. E.g. – *Synchytrium endobioticum.*

When the partial part of the plant body take part in reproduction those fungi are called eucarpic. E.g. – *Saprolegnia sp.*

Q. 16. What is pleomorphic fungi?

Ans. Those fungi having more than one independent form or spore stage in the life cycle are called pleomorphic fungi.

Q. 17. What is amphigynous type of sex organ?

Ans. The oogonium pierces through the antheridium which takes a funnel-like appearence at the oogonial stalk. This type of arrangment of antheridium and oogonium is known as amphigynous type of sex organ.

Q. 18. What is pseudo-paraphyses?

Ans. The asci may or may not be mixed with hyphal structure which are attached both to the roof and to the base of the locule. These hyphal structures having no free ends are known as pseudo-paraphyses.

Q. 19. What is basidiocarp?

Ans. It is the fruit body of Basidiomycetes. It is an umbrella-shaped structure having long massive stalk and broad cap or pileus.

Q. 20. What is fairy ring? Where it is noted?

Ans. At the time of sporulation, sporophore are formed at the tip of the hyphae and thus form a circular ring on the ground known as fairy ring. It is noted in *Agaricus* sp.

Q. 21. What is spermatia?

Ans. In some Ascomycetes e.g. *Neurospora* sp. antheridium is not developed. Specialized sex cells known as spermatia are developed from hyphae.

Q. 22. Name an antibiotics derived from fungi.

Ans. Sir Alexander Fleming in 1928-29 extracted the great antibiotic drug *Penicillin* from *Penicillium notatum.*

Q. 23. What is heterothallism?

Ans. This is a condition where the thalli of a single species are morphologically alike but physiologically different. Physiologically they are unisexual. There is no apparent distinguishing between male and female mycelia. They are different only in their sexual behaviour. This condition of the thallus known as heterothallism.

Q. 24. What is parasexual cycle?

Ans. In some fungi true sexual cycle comprise of nuclear fusion and meiosis is absent. These fungi derive the benefits of sexuality through a cycle known as parasexual cycle. It also defines as a cycle in which plasmogamy, karyogamy and meiosis takes place but not at specific time or at specific points in the life cycle of an organism.

Q. 25. Name an obligate parasite.

Ans. *Puccinia graminis.*

Q. 26. Name an obligate soprohyte.

Ans. *Saprolegnia monoica.*

Q. 27. Name an endophytic mycorrhiza.

Ans. *Rhizotocnia solani.*

Q. 28. What is diplanetism?

Ans. In *Saprolegnia* diplenetism is noted. Diplanetism is the phenomenon of production of two types of zoospore – such zoospores in which two swimming periods occur.

Q. 29. What is monoclinous?

Ans. When antheridia develop on the same hyphae which bears the oogonia then they are said to be monoclinous.

Q. 30. What is diclinous?

Ans. Antheridia may develop terminally on different slender hyphae, generally located near the oogonial hyphae – such antheridia are called diclinous.

Q. 31. What is androgynous?

Ans. When antheridia arise immediately below the oogonium, i.e. from the stalk of the oogonium, then they are called androgynous.

Q. 32. What is dolipore septum?

Ans. In Basidiomycetes, at the junction of two cells i.e. of the septum, there presents a pore-like structure. The pore is perforated by two dumb-bell-shaped structures— that septum is called dolipore septum.

Q. 33. Name different types of fruit body of ascomycetes.

Ans. Apothecium, Perithecium, Cleistothecium.

Q. 34. How does algal cell wall differ from fungal cell wall?

Ans. The algal cell wall is composed of true cellulose, while the fungal cell wall is composed mainly of chitin.

Q. 35. What do you mean by haplobiontic life cycle?

Ans. When a life cycle is dominant by haploid phase, diploid phase is very short and restricted in zygote only. After reduction division the zygote again develops into haploid plant-body. This life cycle is called haplobiontic life cycle.

Q. 36. What is coenozoospore?

Ans. The zoospore of *Vaucheria* contains numerous nuclei and chloroplast in addition to numerous pairs of heterokont flagella, almost equal in length. This type of zoospore is known as coenozoospore.

Q. 37. What is ergotism?

Ans. The sclerotia of the fungus *Cleviceps purpurea,* due to its several poisonous alkaloids contents, are responsable for the physiological disease of domestic animals and man when the sclerotia are consumed by them along with the harvested grain. Such disease is known as ergotism.

Q. 38. Name a lignicolous fungi.

Ans. *Polyporus* sp.

Q. 39. Name one coprophilous fungi.

Ans. *Ascobolus scatigenus.*

Q. 40. What is Buller's phenomenon?

Ans. When a dikaryotic mycelium meets a monokaryotic mycelium the two mycelium form hyphal fusion with one another, and then the dikaryotic mycelium diplodise the monokaryotic mycelium. The diplodisation of mono-karyotic mycelium by dikaryotic mycelium has been called Buller's phenomenon (after Buller who first observed this).

Q. 41. What is paraphysis?

Ans. In the basidiocarp of certain Basidiomycetes fungi contain sterile structures which are associated with basidia in the hymenial layer known as paraphysis.

Q. 42. Give the botanical name of puff-ball fungus.

Ans. *Lycoperdon alveolatum.*

Q. 43. What is ballistospore?

Ans. The term ballistospore has been applied to violently projected spore.

Q. 44. What is clamp connection?

Ans. In majority of Basidiomycetes the secondary mycelium is characterised by the presence of bridge-like hyphal connections known as clamp connection.

Q. 45. What is acervulus?

Ans. Some Deuteromycetes fungi produce small or large cushion-like masses of hyphae with a bed of conidiophore from which conidia are produced – these are known as acervulus.

Q. 46. Name a xerophytic fungi.

Ans. *Xeromyces bisporous.*

Q. 47. What is arbuscule?

Ans. In parasitic fungi the hyphae usually grow inter- or intra-cellularly in the host tissue. The intercellular hyphae send out short organs through host cells known as arbuscule.

Q. 48. Name one obligate biotroph.

Ans. *Puccinia graminis.*

Q. 49. Name one macrocyclic rust.

Ans. *Puccinia graminis.*

Q. 50. What is crozier?

Ans. In class Ascomycetes, at the time of ascus formation, the tip of each ascogenous hyphae bend and form a hook-like structure known as crozier.

Q. 51. What is auxospore? What is its function?

Ans. The progressive dimination in size of some diatom cells does not continue indefinitely because cells of a certain reduced size produce a special type of revenescent cell called auxospore.

It helps in reestablishment of the original size of those smaller cells.

Q. 52. What is ectoparasite?

Ans. When parasitic fungi grow superficially on the surface of the host plant without penetrating the host tissue they are called ectoparasite. E.g. – *Saprolegnia parasitica.*

Q. 53. What is sclerotium?

Ans. After the destruction of ovarian tissue of rye/barley/wheat flower, by the fungus *Cleviceps purpurea* the mycelium of the fungus becomes transformed into dark, hard, club-shaped, pseudo, parenchymatous body called sclerotium.

Q. 54. What is stromata?

Ans. In certain fungi, due to the more or less localised mycelial growth, an accumulated mass of mycelial structure is formed which is known as mycelial mat. But further growth of this mycelial mat may develop into a compact structure called stromata.

Q. 55. Name a fungus of medicinal value and the alkaloid obtained from it.

Ans. *Cleviceps purpurea.* The alkaloids are ergotamine, ergotoxine, ergotoline etc.

Q. 56. Name the primary and secondry hosts in *Puccinia*.

Ans. Primary host – wheat *(Triticum aestivum)*

Secondary host – Barberry *(Berberis vulgaris)*

Q. 57. What do you mean by dimietic basidiocarp?

Ans. In *Polyporus* sp there is found generative hyphae and binding hyphae. When the generative hyphae combine with the binding hyphae then such fruit body are called dimietic basidiocarp.

Q. 58. What are fungicides?

Ans. The chemical substances that cause destruction of fungi are known as fungicides. E.g. – Bavastine.

Q. 59. What is haustoria?

Ans. A haustoria is an organ that is developed from a hyphae usually performing the function of absorption.

Q. 60. What are the differences between dikaryotisation and diplodisation?

Ans. The dikaryotisation is a special type of reproduction process in Basidiomycetes. By this process monokaryotic primary mycelium is converted into dikaryotic secondry mycelium. To some extent this process may be called diplodisation because monokaryotic cells produce dikaryotic cells. In Ascomycetes inside the fruit body the ascogenous hyphae, by the diplodisation process there, develop many dikaryotic cells.

Q. 61. What is an arthospore?

Ans. Arthospore are mycelial fragments. The hyphae of some Basidiomycetes fungi often break up into unicellular unthickened segments called arthospore.

Q. 62. What is hypertrophy?

Ans. It is an abnormal increase in size of one or more organs of a plant in response to the attack of a pathogen.

Q. 63. What is autoecious fungi?

Ans. The parasitic fungi which complete their life cycle on the tissue of a single host are called autoecious fungi.

Q. 64. What is heteroecious fungi?

Ans. The parasitic fungi which complete their life cycle on two hosts are known as heteroecious fungi. E.g. – *Puccinia graminis.*

Q. 65. What is secondary zoospore?

Ans. The zoospore which are kidney-shaped or bean-shaped bearing two oppositely directed whiplash or tinsel type of flagella inserted laterally in the furrow on the concave side are called secondary zoospore. This kind of zoospore develop for primary zoospore.

Q. 66. What is aplanospore?

Ans. These are unicellular, non-motile, non-flagellated spores formed within the sac-like structures. E.g. – *Mucor, Rhizopus* etc.

Q. 67. What is the difference between oospore and zygospore?

Ans. Oospore is the thick-walled resting zygote formed from a fertilised oosphere.

Zygospore is a thick-walled resting spore formed by the fusion of the contents of two gametangia.

Q. 68. Which fungus is called green mould?

Ans. *Penicillium* sp.

Q. 69. What is hypopodium?

Ans. A short branch of one or two cells known as hypopudium may develop on epiphytic mycelium.

Q. 70. Name one bracket fungus.

Ans. *Polyporus* sp.

Q. 71. From where is `penicillin' obtained?

Ans. Penicillin is obtained from fungi *Penicillium notatum.*

Q. 72. What is primary zoospore?

Ans. In some fungi at the time of asexual reproduction there develop pear-shaped zoospores which bear two equally attached flagella of equal length. The zoospores are called primary zoospores.

Q. 73. Name the different sexual spores in *Puccinia graminis.*

Ans. Uredospore, basidiospore, Aeciospore, teliospore, pycniospore.

Q. 74. What is VAM?

Ans. VAM is vasicular arbuscular mycorrhiza.

Q. 75. Where do you find dolipore septum?

Ans. In the class of Basidiomycetes.

Q. 76. What do you mean by plectenchyma?

Ans. Sometime the normal hyphae are so compactly interwoven that the whole mass becomes plate-like – this is called plectenchyma.

Q. 77. Name the types of mycorrhizal relation.

Ans. Ectomycorrhiza, Endomycorrhiza and ecto-endomycorrhiza.

Q. 78. What is ascocarp?

Ans. The fruit body of Ascomycetes fungi is called ascocarp. It is multicellular and remains enveloped by a sheath of sterile hyphae.

Q. 79. What is coprophilous fungi?

Ans. The fungi which mainly grow on animal dung are known as coprophilous fungi. Eg. – *Ascobolus* sp.

Q. 80. What is apothecium?

Ans. The cup shaped or saucer-shaped fruit body of Ascomycetes fungi are known as apothecium.

5 PLANT PATHOLOGY

Q. 1. What is facultative parasite?

Ans. When fungi pass their mode of life as saprophyte in the beginning and later or under certain condition, becomes parasite causing a disease they are known as facultative parasite.

Q. 2. What is macrocyclic rust?

Ans. Macrocyclic is long-cycled life cycle – in this case at least one or more than one type of binucleate spores in addition to telutospores are produced in a complete life cycle.

Q. 3. What is mycorrhiza?

Ans. Sometimes fungi also live in close association with the roots of gymnosperms and angiosperms; this association between fungal hyphae and roots of higher plants is known as mycorrhiza.

Q. 4. What is hypertrophy?

Ans. The most apparent effect is some disease is the over-growth of one or more plant organs or some parts of them and when this phenomenon occurs due to increase of cell size it is called hypertrophy.

Q. 5. What is ectoparasite? Give an example.

Ans. When parasitic fungi grow superficially on the surface of the host plant without penetrating the deeper host tissue it known as ectoparasite.

Q. 6. What is inoculum?

Ans. The portion of the individual pathogen which are brought into contact with the host plant is called inoculum.

Q. 7. Distinguish between pathogen and parasite.

Ans.

Pathogen	*Parasite*
1. It is an agent that by its persistent association causes the disease	1. Parasite absorbs nutrition from the host and ultimately kills them.
2. It may be living or non-living	2. It is generally living.

Q. 8. What are systemic fungicides?

Ans. Systemic fungicides are those which are fungicidal after absorption and subsequent translocation and are able to control the growth of pathogen in the host tissue.

Q. 9. What are phytoalexin?

Ans. Phenolic substances which do not exist in the host tissue but are formed as a result of host-pathogen interaction are known as phytoalexin.

Q. 10. What do you mean by wild fire toxin?

Ans. Wild fire toxin, produced by *Pseudomonas tabaci,* is monotropic, as are many of the toxins produced by bacteria which attack animals. This is an unstable toxin.

Q. 11. Name the toxin which produces HV toxin.

Ans. *Helminthosporium victoriae.*

Q. 12. What is afla toxin?

Ans. It is a type of mycotoxin produced by fungi in fruits, vegetables, foodgrains, seeds, ligume etc. Mainly it is produced from *Aspergillus flavus.*

Q. 13. Define pathotoxin. Give an example.

Ans. A few toxins are species-specific. They only affect plants susceptible to the pathogen i.e. host-specific toxins are called pathotoxin. E.g. Fusaric acid.

Q. 14. Define biopesticides.

Ans. The pesticides which are derived from biological origin.

Q. 15. What do you mean by necrotic symptoms?

Ans. These symptoms are evidence of death of the affected tissue caused by both parasitic and non-parasitic causal agents.

Q. 16. Name the bacterial phytopathogen.

Ans. *Xanthomonas oryzae* cause bacterial blight.

Q. 17. Name one microcyclic heteroceous rust.

Ans. *Puccinia graminis.*

Q. 18. Name three commonly used sulpher fungicides?

Ans. (i) Nabam

(ii) Zinab

(iii) Ferbam etc.

Q. 19. Distinguish between toxin and enzyme.

Ans.

Toxin	*Enzyme*
1. Toxins are biochemical substances produced by pathogens during host infection.	1. Enzymes are metabolic products produced by fungi during metabolic process.
2. These are mainly of three types — zootoxin, phytotoxin and microbial toxin.	2. These are of three types — exoenzyme, endoenzyme, and adaptive enzyme.

Q. 20. Distinguish between ectotrophic and endotrophic mycorrhizae.

Ans.

Ectotrophic	*Endotrophic*
1. In this type many intercellular fungal hyphae occur largely in the cortex and remain connected with well-developed external mycelium.	1. In this type infected root contains well-developed hyphae running in the intercellular spaces and penetrating many of the living cortical cells.
2. It occurs in almost all forest tree.	2. All orchids have this type of mycorrhizae.

Q. 21. What is non-parasitic plant disease?

Ans. When the disease is caused by non-parasitic causal agent it is called non-parasitic plant disease.

Q. 22. What is an alternate host?

Ans. When a pathogen needs two hosts to complete its life cycle then one host is designated as the principal host and the other as alternate host.

Q. 23. What is pathogenesis?

Ans. The chain of metabolical event which brings about the disease is pathogenesis. It involves the action of the pathogen, susceptibility of the plant and impact of the environment.

Q. 24. Name one toxin produced by a plant pathogen.

Ans. Wild fire toxin produced by *Pseudomonas tabaci.*

Q. 25. What is disease triangle?

Ans. In order for disease to occur, three factors must be present. Because of this, disease is often pictured as a triangle having three equal sides. Each side of the disease triangle is necessary in order for disease to occur. The three factors are host, causal agent and environment.

Q. 26. What is a disease cycle?

Ans. In every infectious disease a series of more or less distinct events occur in succession and leads to the development and perpetuation of the disease and the pathogen. This chain of events are called disease cycle.

Q. 27. Define host defense mechanism.

Ans. The mechanism by which the plants protect themselves from the unfavourable environment or from a pathogen is called defense system or defense mechanism.

Q. 28. What is structural defense mechanism?

Ans. To protect the host from the pathogen or environment occurs some structural modifications. These structural modifications act as first line barrier of the plant for pathogen or environmental hazards.

Q. 29. What is PR protein?

Ans. PR protein are pathogen-related protein. They are generally a group of plant protein that are toxic to invading fungal pathogen or other pathogen.

There are many PR proteins – PR_1, PR_4 etc.

Q. 30. What is hypersensitivity?

Ans. The term denotes increased sensitivity as in the rapid death of a host cell in the immediate vicinity at the site of infection.

Q. 31. What is VAM?

Ans. VAM is Vasicular Arbuscular Mycorrhizae. This type of fungi are endomycorrhizal – which invade deeply into the root. It has been observed that VAM are abundanly present in herbaceous plant growing in cultivated soil and grasslands.

Q. 32. What is Bordeaux mixture?

Ans. Bordeaux mixture is a chemical which is used for control of disease. 4 : 4 : 50, 5 : 5 : 50 ratio of copper sulphate, quicklime and water mixture forms Bordeaux solution.

Q. 33. Define suppressive soil.

Ans. Several soil-born pathogens (*Phytopthora* sp) can infect a host much rapidly through live in a particular soil but when they are transferred to other soil the potentiality of those pathogens to form a disease become less. Then the later soil is called supressive for those particular pathogens. It is due to that the soil destructs the inoculum by resident antagonists.

Q. 34. What are biotrophs?

Ans. Parasitic organisms which are deriving their nutrients from living host cell.

Q. 35. What are necrotrophs?

Ans. The organisms which derive their nutrients from cells they have already killed. Most biotrophs cause minimal damage to the host tissue.

Q. 36. What is symbiosis?

Ans. In some cases two living organisms may live non-pathogenically in association with one another. In such cases both organisms are mutually benefited through their life processes. The phenomenon is known as symbiosis.

Q. 37. What do you mean by endemic disease?

Ans. A disease which constantly occurs from year to year in moderate to severe form in a particular area is an endemic disease.

Q. 38. Name a host and a pathogen which produces wild disease in plant.

Ans. Pathogen – *Fusarium udum.*

Host – Pigeon Pea (*Cajanus cajan*).

Q. 39. What are the principles of plant disease control?

Ans. (i) Regulatory

(ii) Cultural

(iii) Biological

(iv) Physical

(v) Chemical methods.

Q. 40. What are antibiotics?

Ans. A chemical compound produced by a microorganism which inhibits or kills other microorganism is known as antibiotic. E.g. Ampicilin, tetracyclin etc.

Q. 41. What is Koch's Postulates?

Ans. When a pathogen is found on a disease plant the pathogen is identified by reference to special manuals which was firstly proposed by Robert Koch. So according to his name the protocol is known as Koch's Postulate.

Q. 42. What is callus?

Ans. A mass of thin-walled undifferenciated cells which develop as the result of culture on nutrient media.

Q. 43. What is appresorium?

Ans. The swollen tip of a hyphae or germ tube that facilitate attachment and penatration of the host by the fungi is called appresorium.

Q. 44. What is canker?

Ans. A necrotic, often sunken lesion on a stem branch or twig of a plant is known as canker.

Q. 45. What do you mean by anthacnose?

Ans. Elongated more or less angular spots appear on the veins on the lower side of the leaf and are known as anthacnose.

Q. 46. What is the control measure of Tobacco mosaic virus?

Ans. (i) Seed bed should be located at a great distance from the tobacco warehouses and free from any tobacco refuse.

(ii) Seed bed soil should be sterilised by steam.

(iii) Diseased plants should be removed and burnt.

(iv) Previous year's plant debris should be destroyed by burning.

Q. 47. What do you mean by obligate parasite?

Ans. Some pathogens cannot complete their life cycles except in a parasitic relation to their respective hosts – they are known as obligate parasites.

Q. 48. What is theraputant?

Ans. It is the chemical material which initiates a curative and mitigating effect on the plant to reduce pathogenic diseases.

Q. 49. Name an antagonistic plant.

Ans. *Asparagus* sp., *Marigolds* sp.

Q. 50. What is incubation period?

Ans. There may be a long period between penetration and infection and appearing of the disease. The interval of time between the penetration and first appearance of symptoms of the host is known as incubation period.

Q. 51. Write the name of the host and pathogen of the disease of Loose smut of wheat.

Ans. Host : *Triticum aestivum*

Pathogen : *Ustilago nuda* ver *tritici.*

Q. 52. Write the name of the host and pathogen of the disease of Late blight of potato.

Ans. Host : *Solanum tuberosum.*

Pathogen : *Phytophthora infestans.*

Q. 53. Write the name of the host and pathogen of the disease of Rust of wheat.

Ans. Host : *Triticum aestivum.*

Pathogen : *Puccinia graminis.*

Q. 54. Write the name of the host and pathogen of the disease of White rust of crucifer.

Ans. Host : Cruciferous crops.

Pathogen : *Albugo candida.*

Q. 55. Write the name of the host and pathogen of the disease of Blast disease of Rice.

Ans. Host : *Oryza sativa.*

Pathogen : *Pyricularia grisea.*

Q. 56. Write the name of the host and pathogen of the disease of Brown spot of rice.

Ans. Host : *Oryza sativa.*

Pathogen : Helminthosporium oryza.

Q. 57. Write the name of the host and pathogen of the disease of Citrus canker.

Ans. Host : *Citrus* sp.

Pathogen : *Xanthomonas citri.*

Q. 58. What is ectomycorrhizae?

Ans. Ectomycorrhiza consist of dense sheath of fungal cell surrounding the root of Angiosperms and Gymonosperms. The fungus often belongs to Basidiomycetes.

Q. 59. What is endomycorrhizae?

Ans. When the ectomycorrhizae penetrate between the cell of the epidermis and the first few cells of the cortical region forming an intercellular hyphae network in the roots of higher plants. It is called endomycorrhizae.

IMPORTANT DISEASES

PLANT DISEASES

(i) Caused by Fungi

No.	Disease	Crop	Causal Organism
1.	Club root	Cabbage	*Plasmodiophora brasicae*
2.	Wart disease	Potato	*Synchytrium endobioticum*
3.	Damping off	Various seedlings	*Pythium* sp.
4.	Late blight	Potato	*Phytophthora infestens*
5.	White rust	Crucifers	*Albugo candida or Cystopus candida*
6.	Downy mildew	Peas	*Peronospora pisi*
7.	Downy mildew	Grapes	*Plasmopara viticola*
8.	Powdery mildew	Grapes	*Uncinula necator*
9.	Powdery mildew	Peas	*Erysiphe polygoni*
10.	Powdery mildew	Wheat	*Erysiphe graminis*
11.	Loose smut	Wheat	*Ustilago nuda tritici*
12.	Covered smut	Barley	*Ustilago hordei*
13.	Loose smut	Barley	*Ustilago nuda*
14.	Covered smut	Oats	*Ustilago kolleri*
15.	Loose smut	Oats	*Ustilago avenae*
16.	Whip smut	Sugarcane	*Ustilago scilaminea*
17.	Grain smut	Jowar	*Sphacelotheca sorghii*
18.	Smut	Bajra	*Tolyposporium penicillariae*
19.	Black rust	Wheat	*Puccinia graminis tritici*
20.	Brown rust	Wheat	*Puccinia recondita*
21.	Yellow rust	Wheat	*Puccinia striiformis*
22.	Rust	Linseed	*Melampsora lini*
23.	Early blight	Potato	*Alternaria solani*
24.	Tikka disease	Groundnut	*Cercospora personata*
25.	Red rot	Sugarcane	*Colletotrichum falcatum*
26.	Flag smut	Wheat	*Urocystis tritici*
27.	Leaf curl	Peach	*Taphrina deformans*

(ii) Caused by Viruses

No.	Disease	Causal Virus
1.	Tobacco Mosaic	Tobacco mosaic virus
2.	Abutilon Mosaic	Abutilon mosaic virus
3.	Papaya mosaic	Papaya mosaic virus
4.	Cucumber mosaic	Cucumber mosaic virus
5.	Potato mild mosaic	Potato virus X
6.	Potato rugose mosaic	Potato virus X, Potato virus Y
7.	Potato leaf roll	Potato leaf curl virus
8.	Tomato leaf curl	Tobacco leaf curl virus
9.	Little leaf of brinjal	Brinjal little leaf virus
10.	Little leaf of cotton	Cotton little leaf virus
11.	Ratoon stunt of sugarcane	Ratoon stunt virus
12.	Rosette of groundnut	Groundnut mosaic virus
13.	Tristeza of citrus	Citrus tristeza virus
14.	Bunchy top of banana	Banana bunchy top virus
15.	Sugarcane mosaic	Sugarcane virus I

(iii) Caused by bacteria

No.	Disease and Host	Causal Organism
1.	Wilt of potato	*Pseudomonas solanacearum*
2.	Wilt of tomato	*Pseudomonas solanacearum*
3.	Wilt of tobacco	*Pseudomonas solanacearum*
4.	Wilt of pepper	*Pseudomonas solanacearum*
5.	Blight of beans	*Xanthomonas phaseoli*
6.	Black root of crucifers	*Xanthomonas campestris*
7.	Citrus canker	*Xanthomonas citri*
8.	Ring rot of potato	*Corynebacterium sepidonicum*
9.	Canker of tomato	*Corynebacterium michiganese*
10.	Crown gall of *Rubus* sp.	*Agrobacterium tumefaciens*
11.	Crown gall of apple	*Agrobacterium tumefaciens*
12.	Crown gall of rose	*Agrobacterium tumefaciens*
13.	Crown gall of tomato	*Agrobacterium tumefaciens*
14.	Scab of potato	*Streptomyces scabies*
15.	Fire blight of pear & apple	*Erwinia amylovora*
16.	Soft rot of vegetable	*Erwinia carotovora*
17.	Wilt of cucurbits	*Erwinia tracheiphila*

6 BRYOPHYTA

Q. 1. What is gemma cup?

Ans. The most common method of the vegetative reproduction in some bryophytes take place by means of especially designed vegetative bodies, the gemma, which originated from the cup-like structure, the gemma cup, found on the dorsal surface of the thallus or at the rhizoidal portion. E.g. *Marchantia* sp.

Q. 2. What are elaters?

Ans. In some bryophytes like *Marchentia,* within capsule are found some exceptionally long, slender sterile cells known as elaters. They have double spiral thickening; the mature elaters lack protoplasm and function in spore dispersal.

Q. 3. What is air channel?

Ans. In *Riccia's* photosynthetic zone there present green or chlorenchymatous cells which are arranged in vertical rows or columns. Between the columns of green cells very narrow, deep, vertical slits are formed. These slits are called air channel.

Q. 4. What is nurse cells?

Ans. Pagan (1932) reported the sterility of some spore mother cell in *Riccia crystalina,* such sterile cells are called nurse cells.

Q. 5. What is blepharoplast?

Ans. *Riccia's* androcyte possesses a big prominent nucleus and an extra nuclear granule. This granule is called blepheroplast. Two flagella arise from the anterior end of the blepheroplast.

Q. 6. What do you mean by involucre?

Ans. In *Pellia* archegonia develop on thallus in a group. A delicate membrane arising from the top of the ridge, covers the archegonia as in a pocket. This membranous cover is known as involucre.

Q. 7. What is amphigastria?

Ans. In incubous arrangment of leaves of stem of *Porella* – the small leaves of the ventral row are amphigastria.

Q. 8. What do you mean by incubous arrangement?

Ans. The *Porella's* stem bears three rows of leaf. There are two rows of dorsal leaf and one row of ventral leaf. The dorsal leaves closely overlap each other and this way they cover the stem from above. The anterior edge of the leaf cover the posterior edge of the leaf next above it, when seen from the above. Such an overlapping arrangment of the leaves is knows as incubous arrangement.

Q. 9. What is protonema?

Ans. In bryophytes capsule spore germinates under suitable condition to give rise to main gametophyte plant directly or firstly to a juvenile, filamentous stage called protonema from which later arise the main gametophyte.

Q. 10. Why bryophytes are amphibians of plant kingdom?

Ans. Bryophytes grow in water, land and swamps, the area where the water and land meet. So this group is called amphibian non-vascular group of plant.

Q. 11. What do you mean by juvenile stage?

Ans. In leafy gametophore the meiospore germinates and forms a slender, green branching system of filaments called protonema. This stage is called juvenile stage. The protonema further forms the gametophyte.

Q. 12. Name an epiphytic bryophytes.

Ans. *Dendroceros* sp.

Q. 13. Name a *Riccia's* species where scales are absent.

Ans. *Riccia crystallina.*

Q. 14. Name the bryophyte which grows in the highest altitude.

Ans. A moss *Aongstroemia julacea* has been recorded from an altitude of 19,800 ft of Himalayas.

Q. 15. Name a saprophyte bryophyte.

Ans. *Buxbaumia aphylla* grow upon rotted wood.

Q. 16. Name a fossil bryopsida.

Ans. *Muscites bertrandi.*

Q. 17. Name a monoceious and dioceious species of *Riccia.*

Ans. Monoccious species – *Riccia crystallina.*

Dioceious species – *Riccia discolor.*

Q. 18. Which bryophyte has lowest chromosome number?

Ans. *Takakia* sp (2n = 8).

Q. 19. What is Madotheca?

Ans. *Porella* is known as Madotheca.

Q. 20. Name the aquatic species of bryophytes.

Ans. *Riccia fluitans, Ricciocarpus natans.*

Q. 21. Where peristome teeth are found?

Ans. *Funaria, Polytrichum.*

Q. 22. Name a moss where peristome teeth are absent.

Ans. *Sphagnum* sp.

Q. 23. What is calyptra?

Ans. In bryophytes the embryo remains within the venter of the female reproductive structure, i.e. archegonium. The venter wall enlarges along with the developing embryo to form the protective envelope known as calyptra.

Q. 24. What is elaterophore?

Ans. In some bryophytes, inside the capsule wall are present elaterophore which consisted of a bundle of 20 to 100 stout, erect, fixed elaters. The lower end of the cell of elaterophore remain attach to the floor of the capsule cavity. E.g. *Pellia.*

Q. 25. In which bryophytes, are present parasitic algae?

Ans. In intercellular mauilage cavity of gametophytic plant body of *Anthoceros* are found Cyanophycean algae *Nostoc.*

Q. 26. Write down the homothallic and heterothallic species of *Anthoceros.*

Ans. Homothallic – *Anthoceros fusiformis.*

Heterothallic – *Anthoceros himalayensis.*

Q. 27. What is pseudoelaters? Where do you find it?

Ans. In sporogenous tissue of *Anthoceros's* sporophyte, the sterile spore mother cells soon divide obliqely or transversly producing three- to five-celled structures called pseudoelater. It is found in *Anthoceros.*

Q. 28. What is gametophore?

Ans. The leafy bryophyte in which the prostrate portion contains stem and leaves. The plant body is called gametrophore.

Q. 29. Which sporophyte is the advanced one?

Ans. *Sphagnum.*

Q. 30. What is bulbils?

Ans. These are resting buds developed on the rhizoids of *Funaria* sp.

Q. 31. Name a monoecious and dioecious species of *Riccia.*

Ans. *Riccia crystallina* – monoecious.

Riccia discolor – dioecious.

Q. 32. What is antheridiophore and archogoniophore?

Ans. In *Marchentia* the sex organ are antheridia and archegonia which are born upon the special erect stalked branches called the antheridiophore and archegoniophore respectively.

Q. 33. What is areolae?

Ans. In *Marchantia's* gametophytic plant body the upper surface of the thallus is marked by rhomboidal to polygonal areas called areolae.

Q. 34. What is perigynium or pseudoperianth?

Ans. In *Marchantia* the archegonium base of the venter is developed a collar-like structure. It consist of a ring of cells and surround the archegonial stalk. It is called perigynium or pseudoperianth.

Q. 35. What is spore dispersal mechanism of *Riccia's* sporophyte?

Ans. In *Riccia,* the sporophyte lack of any special mechanism for spore dispersal. The death and decay of both calyptra and also of gametophyte causes dispersal of spore.

Q. 36. Which sporophyte is the simplest among the bryophytes?

Ans. *Riccia* sp. which only consists of capsule.

Q. 37. Where pseudoelater are found?

Ans. Sporophytic plant body of *Anthoceros* sp.

Q. 38. What is spermatids?

Ans. In antheridium of bryophytes each androcyte mother cell has a denser cytoplasm and a relatively larger nucleus. It divides longitudinally to form two sperm cells known as spermatids.

Q. 39. Name a *Porella* species containing gammae.

Ans. *Porella ratundifolia.*

Q. 40. What is parthenogenesis?

Ans. Development of sporophytic plant body from the egg of oogonium without fertilisation is called parthenogenesis.

Q. 41. Which bryophytes are called 'horn worts' and why?

Ans. In *Anthoceros,* the capsule arises from the thallie in the form of small horny structure, they range from two to fifteen centimeters in height. Thus it is known as 'horn worts'.

Q. 42. In which bryophyte the sporophytic plant body consists of only foot and capsule?

Ans. *Anthoceros* sp.

Q. 43. Where are nurse cells present?

Ans. In Sporophyte of *Riccia* sp.

Q. 44. What is archesporium?

Ans. In bryophyte at the time of development of sporophytic plant body the inner layer of amphithecium forms 2-4 layers thick-walled cells call archesporium.

Q. 45. Mention the type of rhizoids of *Marchantia*.

Ans. In *Marchantia* there are found two type of rhizoids. These are

(i) Tuberculate, and

(ii) Smooth-walled.

Q. 46. State the pteridophytic features of *Anthoceros*.

Ans. (i) Present of central sterile columella.

(ii) The elaborate photosynthetic tissue with stomata.

(iii) Green cylindrical sporophyte.

Q. 47. Name one aquatic moss.

Ans. *Sphagnum* sp.

Q. 48. What is retort cells?

Ans. In a few species of bryophytes, some outer cortical cells enlarge peculiarly and appear like retorts – these are known as retort cells.

Q. 49. What is the origin of sporogenous cells in *Anthoceros?*

Ans. The origin of sporogenous cells in *Anthoceros* is endothecium tissue.

Q. 50. How the peristome teeth are arranged in Diplolepideae?

Ans. In Diplolepideae the peristome teeth are arranged in two rows. The outer circle is one ring of 16 layered brown peristome teeth. The inner circle has another 16 peristome teeth in paler colour.

Q. 51. What is rhomboidal cell? Where do you find it?

Ans. In some bryophytes at the midrib portion of some leaves are present some rectangular cells which contain chloroplast – these are called rhomboidal cell. It is found in *Funaria* sp.

Q. 52. Where do thalloid and filamentous protonema occur?

Ans. Thalloid protonema occur in *Sphagnum* sp.

Filamentous protonema occur in *Funaria* sp.

Q. 53. State the economic importance of *Sphagnum*.

Ans. (i) *Sphagnum* used in household purpose.

(ii) It is used as food by animal, bird and also by human as salad.

(iii) It is used as a medicine.

Q. 54. What is peristome teeth? Mention its function.

Ans. In some bryophytes there are curved, teeth-like, thickened fibrous cells situated just above the diaphragm of the capsule covered by operculum – these are known as peristome teeth.

It helps in the spore dispersal. The inner peristome acts as a sieve allowing the spores to escape gradually.

Q. 55. What is an amphithecium?

Ans. Each of the outer cell of young sporogonium is undifferentiated spherical embryo and divides by periclinal wall. A single layer of outer cells is separated from an inner central mass of the cell. The single layer of outer cell is called an amphithecium.

Q. 56. How does succubus type of leaf arrangement differ from the incubous type?

Ans.

Succubus	*Incubous*
1. The posterior margin of each leaf overlies the forward edge of the next older leaf on the same side in dorsal view E.g. *Plagiochila*	1. The overlapping is reversed in the incubous arrangement of leaves. The posterior margin of each leaf underlies the forward edge of the next older leaf in dorsal view. E.g. *Porella*

Q. 57. How do the morphological nature of the leaves of *Sphagnum* and *Funaria* differ?

Ans. The leaves of the *Sphagnum* are small. The nature of the leaves is without midrib entire and shows two different types of cell (i) narrow, living and chlorophyll containing cell (ii) dead, rhomboidal hyaline cell with pores and spiral thickenings.

The leaves of *Funaria* provide a distinct midrib. The leaves are simple, sessile more or less oval and with pointed apex and mature leaves are arranged spirally on the aerial stem.

Q. 58. State the pteridophytic features of *Anthoceros.*

Ans. *Anthoceros* shows sunken archegonium and development of antheridium from hypodermal initial cell-like pteridophyte.

Q. 59. What is pseudopodium?

Ans. In *Sphagnum's* sporophyte seta subpresses in between capsule and foot. Instead of that the sporophyte is embedded in leafless, stalk-like gametophytic plant body. This structure is called pseudopodium.

Q. 60. Which moss is called as \`peat' or \`bog' moss and why?

Ans. *Sphagnum* genus is called peat or bog because they grow in extensive masses on boggy and peaty soil and also as submerged aquatics in peaty pools.

Q. 61. Name a marine water bryophyte.

Ans. *Scapania undulata.*

Q. 62. What are the difference between primary and secondry protonema?

Ans.

Primary protonema	*Secondry protonema*
1. It results from the germinating meiospore. When fully grown it consists of a slender green, branching system of filamentous structure called primary protonema.	1. In mosses, any cell of a detached, injured portion of the stem leaf and rhizoids, when surrounded by moist air is then modified into filamentous protonema. It is known as secondry protonema.

Q. 63. Name an edible bryophyte.

Ans. *Sphagnum fimbriatum.*

Q. 64. Name a medicinally important bryophyte.

Ans. *Polytrichum commune.*

Q. 65. Name an air pollution indicator among among bryophyte.

Ans. *Bryum rubrum.*

Q. 66. Name an Indian species of *Porella.*

Ans. *Porella platyphylla.*

Q. 67. What is marsupium?

Ans. In some leafing Jungerminales the tip of the female branch becomes enlarged into a pouch-like structure. It is called the marsupium.

Q. 68. Name a *Porella* species containing gemmae.

Ans. *Porella ratundifolia.*

Q. 69. Name the sterile cell found within the sporophytes of *Riccia.*

Ans. Nurse cell.

Q. 70. Write the nature of peristome teeth in *Funaria.*

Ans. The peristome teeth are differentiated into two rings - exostome and endostome. There are 16 teeth in each ring. The teeth of the outer series are curved, narrow, triangular plates – thick, brownish red. The teeth of the inner series are comparatively smaller, delicate and colourless.

Q. 71. Name a bryophyte where assimilatory filaments occur.

Ans. *Marchentia* sp.

Q. 72. Write the nature of *Sphagnum* protonema.

Ans. The protonema forming a prostrate green irregularly lobed thalloid structure. It is one cell in thickness and is attached to the substratum by multicellular rhizoids. On this protonema a single bud is developed from a marginal cell and this develops the new leafy gametophore.

Q. 73. State the biological significance of *Anthoceros* sporophyte.

Ans. (i) Ventilated photosynthetic system in the capsule wall that can make much of its own food.

(ii) The central core of sterile tissue forming the columella. It plays a definite role in conduction and affords mechanical support.

Q. 74. What is the function of rhizoids?

Ans. The main function of the rhizoids is to anchor the plant to the substratum. The absorption of water takes place directly through the leaf and stem cell.

Q. 75. What is receptacle?

Ans. In *Pellia* the archegonia are produced in groups on the upper surface of the thallus lobes, just behind the apical cell. All the archegonia in the cluster are found to be situated on a slightly raised transverse ridge of tissue known as receptacle.

Q. 76. Where are multicellular and jacket sex organ present?

Ans. *Funaria* sp.

7 PTERIDOPHYTA

Q. 1. What is ligule?

Ans. In the class Lycopsida each leaf shows a tongue-like structure, called ligule developing from the base of each adaxial surface.

Q. 2. What is trabaculae?

Ans. The mature stem of *Selaginella* endodermis is modified into elongated axis radial to the stele and also laterally separated from one another by big air gaps — this structure is called trabaculae.

Q. 3. What is glossopodium?

Ans. At the base of the ligule there is a hemispherical mass of thin-walled cell known as glossopodium.

Q. 4. What is sporophore?

Ans. In sporocarp of *Marsilea* next to the hypodermal layer there is a layer of paranchymatous cells. These *Marsilea* cells are gelatinised in a mature sporocarp and help in its dehiscence. This gelatinous layer is called sporophore.

Q. 5. What is sporocarp?

Ans. In bisporangiate, heterosporous *Marsilea,* the micro- and mega-sporangia are born in special bean-like bodies called sporocarp.

Q. 6. What is heterospory?

Ans. The phenomenon of production of two types of spore in an individual species of pteridophyte is called heterospory. E.g. — *Selaginella, Marsilea.*

Q. 7. What are the differences between eusporangium and leptosporangium?

Ans.

Eusporangium	*Leptosporangium*
1. The sporangium is larger and massive and the wall is several layers thick.	1. The sporangium is small and the wall is only one layered.
2. Spore content is high.	2. Spore content is low.
3. The adjacent cells are involved in the formation of part of sporangium wall and stalk.	3. Sporangium wall and the stalk as well as the spore are derived from the outer cell.
4. Primitive.	4. Advanced.

Q. 8. What are the xerophytic features of *Equisetum*?

Ans. (i) Presence of thick coat of cuticle all over the epidermal layer of stem.

(ii) Deep-seated rhizome.

(iii) Presence of small scale-like leaves.

(iv) Presence of sunken stomata.

Q. 9. Why is *Lepidocarpon* considered as pseudoseed?

Ans. In *Lepidocarpon* the megasporophyll tissue grows up and develop a false integument which is open at the tip by a long slit forming a false micropyle. On the whole the structure looks like a seed but, because of the absence of integument, absence of any siphonogamy, and absence of a dormant seed stage, it is considered as pseudoseed.

Q. 10. What is carinal canal?

Ans. In *Equisetum* the vascular bundles are arranged in a ring. Each bundle has three small strands of xylem, and inner strand of protoxylem that lies on the side of a cavity. This cavity is known as carinal canal.

Q. 11. Define tapatum.

Ans. A nutritive layer formed partly by the inner layer of the jacket and partly by some outer sporogenous cell of sporangium of *Lycopodium*.

Q. 12. What is incipient heterospory?

Ans. In *Equisetum* homosporous form of spores produce two types of gametophyte—smaller one male and larger one female. The latter becomes hermaphrodite if fertilisation is delayed. No hermaphrodite, in a population the proportion of male and female gametophyte is influenced by environmental condition.

Q. 13. Which fern has walking habit? Why did it called so?

Ans. *Adiantum caudatum* has walking habit.

In *Adiantum*, the pinnae are stalked. The rachis may terminate in a pinnae or may develop and bear a vegetative bud at the tip. Under favourable condition this vegetative bud may develop into a daughter plant which, in turns, repeats the process – leading to the well-known 'walking habit'.

Q. 14. What do you mean by horns or teeth of sporocarp?

Ans. Just beyond the point of attachment of the sporocarp to the stalk there is often present one or two teeth-like protruberances, they are known as horn or teeth of sporocarp.

Q. 15. What is resurrection plant?

Ans. Some xerophytic plants remain folded in a ball-like structure in dry condition. But as soon as they come in contact of water they become unfolded and normal. They are called ressurrection plants. E.g. – *Selaginella lepidotum.*

Q. 16. Give a few example of the occurrence of multiflagellate spermatozoid in pteridophyta.

Ans. *Isoetes, Psilotum.*

Q. 17. What is protocorn?

Ans. In *Lycopodium cernuum* embryo develops in a massive globose structure called the protocorn which becomes green and develops rhizoids below.

Q. 18. What is microphyllus?

Ans. The leaves are simple, sessile and small in size and possess a single median vein that fails to reach to the leaf apex. Such leaves are called microphyllus.

Q. 19. What do you mean by pseudomonopodial branching?

Ans. The two branches of dichotomy stem are unequal – one of them remain small and the other grows to a greater length. The smaller or the weaker branch may ultimately stop growth and bear strobilus. This type of branching is called pseudomonopodial branching.

Q. 20. What are the differences between sorus and synangium?

Ans. Several sporangia are born in a sorus, thus a sorus is a sporangia-containing structure whereas a synangium is a three-lobed sporangium inside which are produced several spore mother cells.

Sorus is found in *Adiantum* sp.

Synangium is found in *Psilotum* sp.

Q. 21. An aquatic specis of fern used in Agriculture.

Ans. *Marsilea* sp.

Q. 22. Distinguish between megasporophyll and microsporophyll.

Ans.

Megasporophyll	*Microsporophyll*
1. Larger in size	1. Smaller in size.
2. Midvein slighly bent and enters inside the leaves.	2. Midvein straighly enters inside the leaf.
3. Midvein reach the apex.	3. Midvein does not reach the apex.

Q. 23. Write the generic name of spike moss.

Ans. *Selaginella* sp.

Q. 24. What is velum? Where do you find it?

Ans. The sporangium is sunken i.e., lies within a depression in the leaf base and is partly or fully roofed over by a membranous outgrowth called `velum', arising just below the ligule. E.g. – *Isoetes* sp.

Q. 25. In which geological age did the *Calamites* disappear?

Ans. The *Calamites* disappeared early in the Triassic.

Q. 26. What is bisporangiate sorus?

Ans. When the sorus contains two types of sporangia – i.e. megasporangia and microsporangia – then this type of sorus is known as bisporangiate sorus. E.g. *Marsilea minuta.*

Q. 27. Give a account of the geological distribution of *Zosterophyllum.*

Ans. *Zosterophyllum myretonium* explored from the lower old red sandstone of Scotland. *Zosterophyllum* fossils have been obtained from the upper Silurian to lower Devonian.

Q. 28. What is the present name of *Rhynia major*?

Ans. *Aglaophyton major.*

Q. 29. What is the morphological nature of the cone of *Equisetum*?

Ans. The morphological nature of the cone of *Equisetum* is peculiar. It consits of a central axis which bears a number of densely crowded T-shaped peltate discs called sporangiophore.

Q. 30. What is strobilus?

Ans. *Lycopodium* and *Selaginella* present a leaf-like structure. Some botanists regarded them as spore-bearing leaves. These true cone structures are called strobilus.

Q. 31. Name two heterosporus Lycopod.

Ans. *Selaginella monospora*

Isoetes coromondelina.

Q. 32. Name the vestigial organ found in pteridophyte.

Ans. (i) Elaters – *Equisetum* spore.

(ii) Annulus – *Equisetum* cone.

Q. 33. Name a pendulous epiphytic species of *Psilotum.*

Ans. *Psilotum flaccidum.*

Q. 34. What is perichnos?

Ans. Perichnos are longitudenal channels of thin-walled loose tissue transversing the length of the leaf parallel to the vein. Perichnos serve the purpose of aeration. It is observed in the genus *Lepidocarpon.*

Q. 35. Name a pteridophyte where elaters are found.

Ans. *Equisetum* sp.

Q. 36. Name a fossil genus where you find ligule.

Ans. *Lepidocarpon* sp.

Q. 37. What are the fern characteristic of *Marsilea*?

Ans. (i) Rhizomatous stem

(ii) Compound foliage leaves

(iii) Circinate venation of leaf

(iv) Occurence of micro- and mega-sporangia in sori covered with indusium.

Q. 38. Name a cone genus of *Calamites*.

Ans. *Calamostachya* sp.

Q. 39. In which fern circinate venation is absent?

Ans. *Ophioglossum* sp.

Q. 40. Where do you get mixed pith in Pteridophyte?

Ans. *Adiantum* sp.

Q. 41. What is raphe? Where do you find it?

Ans. The fused portion of the base of the sporocarp and the end portion of the pedicel is called raphe. It is found in *Marsilea* sp.

Q. 42. Name one heterosporous fossil Pteridophyte.

Ans. *Lepidodendron*.

Q. 43. Mention two difference between the t.s. of stem of *Calamites* and *Equisetum*.

Ans. (i) *Equisetum* have a secondary thickening by the activity of cambium, whereas in *Calamites* there is no secondary wood.

(ii) Valecular canals were absent in *Calamites* but present in *Equisetum*.

Q. 44. What type of stele is found in *Rhynia's* stem?

Ans. Protostele (Haplostele). The central xylem core is completely surrounded by phloem layer which is 4 to 5 cells in thickness.

Q. 45. How many species include the genus *Zosterophyllum*?

Ans. There are 3 species :

(i) *Zosterophyllum myretonianum*.

(ii) *Zosterophyllum australionum*.

(iii) *Zosterophyllum rhonanum*.

Q. 46. Name one economic valuable species of *Lycopodium*.

Ans. *Lycopodium* powder is a commercial product used in medicine as well as in pyrotechnic.

Q. 47. Where do you find actinostele, mixed protostele and plectostele?

Ans. Actinostele – stem of *Lycopodium selago*.

Mixed protostele – stem of *Lycopodium cernuum*.

Plectostele – stem of *Lycopodium clavatum*.

Q. 48. What type of reproductive structure is found in genus *Lycopodium*?

Ans. Spike-like strobile compactly covered by sporophylls homosporous.

Q. 49. Which Pteridophyte is named `club moss'?

Ans. *Lycopodium* sp. have moss-like appearance in club-shaped strobile earn them the name club moss.

Q. 50. Name one epiphytic *Lycopodium* species.

Ans. *Lycopodium phlegmaria.*

Q. 51. Name a Pteridophyte where you find gemmae.

Ans. *Lycopodium selago.*

Q. 52. Name a xerophytic species of *Selaginella.*

Ans. *Selaginella proniflora.*

Q. 53. Where do you find trabeculae?

Ans. *Selaginella kraussiana.*

Q. 54. Where do you find a single megaspore in megasporangium of *Selaginella*?

Ans. *Selaginella monospora.*

Q. 55. Name a climbing species of *Selaginella.*

Ans. *Selaginella alligans.*

Q. 56. Which Pteridophyte is commonly known as `ground pine'?

Ans. Species of *Lycopodium.*

Q. 57. What are the names of the leaves and strobilus from genus *Lepidodendron*?

Ans. Leaves – *Lepidophyllum*

Strobilli – *Lepidostrobus*

Q. 58. Write the name of the underground axes and leaf base of genus *Lepidodendron.*

Ans. Underground axes – *Stigmaria*

Leaf base – *Lepidophloids.*

Q. 59. Which Pteridophyte is commonly known as `horsetailed scouring rushes and why?

Ans. *Equisetum* sp.

The aerial part of *Equisetum* divides into branches. Reproductive shoots, which are looked like horse tail so this pteridophyte is called as horse tailed scouring rushes.

Q. 60. Name an Indian species of *Equisetum.*

Ans. *Equisetum debile.*

Q. 61. Name the smallest and largest species of *Equisetum.*

Ans. Smallest – *Equisetum scripoids*

Largest – *Equisetum gigantum.*

Q. 62. What are the root and leaf genus of *Calamites?*

Ans. Root – *Asteroxyelon*

Leaf – *Annularia, Asterophyllites.*

Q. 63. Name an Indian species of *Marsilea.*

Ans. *Marsilea quadrifolia.*

Q. 64. Name a xerophytic *Marsilea* species.

Ans. *Marsilea condensata.*

Q. 65. Name an edible species of *Marsilea.*

Ans. *Marsilea drummondii* full of starch is used as food in Australia.

Q. 66. Which Pteridophyte is commonly known as `Maiden-hair fern'.

Ans. *Adiantum* sp.

Q. 67. Name a Indian species of *Adiantum.*

Ans. *Adiantum caudatum.*

Q. 68. In which Pteridophyte will you find false indesium?

Ans. In sporophyll of *Adiantum* sp.

Q. 69. In which period is *Archaeopteris* found?

Ans. *Archaeopteris* found in Middle Devonian period.

Q. 70. Name a heterosporous species of *Archaeopteris.*

Ans. *Archaeopteris latifolia.*

Q. 71. Which Pteridophyte is commonly known as `whisk fern'?

Ans. The genus of *Psilotum.*

Q. 72. Name a rootless pteridophyte.

Ans. *Psilotum* sp.

Q. 73. Name a homophyllum and heterophyllum species of *Selaginella.*

Ans. Homophyllum – *Selaginella rupestris*
Heterophyllum – *Selaginella kraussiana.*

Q. 74. Name a creeping species of *Adiantum.*

Ans. *Adiantum capillus – veneris.*

Q. 75. Name a species of *Lycopodium* where protocorn develops.

Ans. *Lycopodium cernuum.*

Q. 76. What is exoscopic embryogeny?

Ans. In exoscopic embryogeny the apex of the embryo is directed toward the neck of the archegonia. E.g. – *Psilotum* sp.

Q. 77. Mention the salient characters of *Zosterophyllum.*

Ans. (i) Sporangia are born on very short lateral stalks spirally arranged on the stem.

(ii) The aerial stem is leafless.

(iii) The sporangia are rather wide and each is provided with an apical slit for spore dispersal.

Q. 78. Mention the important similarities of *Selaginella* and Gymnosperms.

Ans. (i) *Selaginella* and Gymnosperms are both heterosporous.

(ii) Both possess independent sporophyte that are differentiated into root, stem and leaves.

Q. 79. Name the cone genus of *Calamites*.

Ans. *Palaeostachya.*

Q. 80. Name a species of *Lycopodium* having sporophytic prothallus.

Ans. *Lycopodium phlagmaria.*

Q. 81. Name one species of *Lycopodium* belonging to the sub genus *Urostachya*.

Ans. *Lycopodium selago.*

Q. 82. Who proposed the telome concept?

Ans. Zimmermann (1930).

Q. 83. How many species are present in the genus *Psilophyton*?

Ans. **4 distinct species.**

8 GYMNOSPERM

Q. 1. What is platyspermic seed? Give one example.

Ans. The bilaterally symmetrical seed is called platyspermic seed.

E.g. *Rhabdosperrmum, Cardainthus.*

Q. 2. Mention the xerophytic features of *Ephedra.*

Ans. (i) The epidermis consist of thick-walled cells with thick cuticle.

(ii) Stomata are sunken.

Q. 3. How does the nutritive tissue of gymnospemous embryo differ from that of angiospermous one?

Ans. (i) The nutritive tissue i.e. endosperm is formed after fertilisation among the angiosperm, whereas in gymnosperm it is formed before fertilisation.

(ii) The endosperm is triploid among angiosperms but it is haploid in gymnosperms.

Q. 4. What is meant by `radiospermic' ovule? Give one example.

Ans. Radially symmetrical ovules are called radiospermic ovule. E.g. *Cardiocarpus magnicellularis.*

Q. 5. Name the gymnospermous source of turpentine.

Ans. *Pinus insularis.*

Q. 6. What is rosette polyembryony?

Ans. In *Pinus* at the time of embryogenesis 4 nuclei derived from zygote divides again and arrange themselve in two tiers. These two tiers form four tiers (i) embryonal tier (ii) suspensor tier (iii) rossette tier (iv) upper tier. Sometimes the cell of rossette tier also develop into embryo and this is called rossette polyembryony.

Q. 7. Mention the economic importance of *Ephedra.*

Ans. *Ephedra* produces a medicine called Ephedrine which is used against asthma and bronchial infections.

Q. 8. What is meant by girdling of leaf traces? Where does it occur?

Ans. A leaf trace, after arising from the stellar cylinder, usually does not pass into the nearest leaf directly but turns round a semicircle or `girdles' the stem before passing to the next higher leaf. It is called girdling of leaf traces. E.g. *Cycas* stem.

Q. 9. Name a Linean species of *Gnetum*.

Ans. *Gnetum ula.*

Q. 10. Name a leaf genus of extinct gymnosperm.

Ans. *Glossopteris.*

Q. 11. Write two angiospermic characteristics of *Gnetum*.

Ans. (i) Presence of perianth in male and female flowers of *Gnetum* resembles angiosperms.

(ii) Like dicot leaves, the leaves of *Gnetum* are broad with reticulate venation and arranged in opposite decussate manner.

Q. 12. What is meant by the term trachaeophyte?

Ans. Pteridophyta, gymnosperm and angiosperm are called trachaeophyte i.e. vascular plants due to the presence of vascular tissue in their sporophytic plant body mainly.

Q. 13. Name a gymnosperm where you find endophytic Cyanobacteria.

Ans. *Cycus circinalis* is a gymnosperm where endophytic Cyanobacteria is present.

Q. 14. Write the botanical name of maiden hair tree.

Ans. *Ginkgo biloba.*

Q. 15. Name a common species of *Pinus* occuring in Western Himalaya.

Ans. *Pinus roxburgii.*

Q. 16. Name the female fructification of *Glossopteris*.

Ans. *Scutum, Pluma* are the female fructification of *Glossopteris.*

Q. 17. What is albuminous cell?

Ans. In the vascular bundle of *Pinus* leaflet some phlem parenchymatous cells are rich in dense cytoplasm – such cells are known as albuminous cell.

Q. 18. Name a tree species of *Gnetum*.

Ans. *Gnetum gnemon.*

Q.19. What is sulphur shower?

Ans. In *Pinus* sp the pollen grain are winged and yellow in colour. When mature, each microsporangium dehisce by a longitudinal slit along the long axis. As a result a large number of pollen grains are released forming a cloud which is often called `sulphur shower'.

Q. 20. Mention the age and Indian locality of *Glossopteris*.

Ans. Age – Lower Gondwana.

Locality – Barakar and Raniganj of the Damodar series.

Q. 21. Mention two important features which differenciate Cycadophyta from Coniferophyta.

Ans.

Cycadophyta	*Coniferophyta*
1. Here monoxylic wood is found.	1. Here pycnoxylic wood is found.
2. It has wide medullary ray.	2. It has narrow medullary ray.

Q. 22. Where do you find endosperm as a post fertilisation structure in gymnosperm?

Ans. *Gnetum.*

Q. 23. What is the nature of venation in *Glossopteris*?

Ans. There was a very strong midrib from which numerous fine veins pass out to the margin forming a network i.e. reticulate venation.

Q. 24. Name two gymnospermous genera where cilliated male gametes are formed.

Ans. *Cycas* sp. *Ginkgo* sp.

Q. 25. What is tentpole?

Ans. The archegonial chamber of *Ginkgo* is cylindrical sorrounding a columnar mass of gametophytic tissue upon which the nucellar tissue rests like a tent of pole – hence the columnar gametophytic tissue is called tent pole.

Q. 26. What is cleavage polyembryony?

Ans. In the conifers all the four cells of the young embryo separate and develop into 4 embryos but only one completely developed and others are aborted. This type of polyembryony is called cleavage polyembryony.

Q. 27. Name the place where *Williamsonia* is found in India.

Ans. *Williamsonia* is found in the Upper Gondwana Rajmahal and Jabalpur series.

Q. 28. What is `Bars of Sanio'? Where do you find it?

Ans. Some gymnosperms have small trabeculae extending across the lumen called `Bars of Sanio'. These trabeculae are made of cellulose. It is found in *Arucaria* sp.

Q.29. What is interseminal scale? Where does it occur?

Ans. It is a sterile structure. Sometime in some gymnosperm the megasporophyll ovule lie intermingled with a sterile structure known as interseminal scale. It occur in *Cycadocidea.*

Q. 30. Mention the diagnostic features of *Glossopteris* leaf.

Ans. *Glossopteris* leaves are simple, entire, sessile or very shortly petioled linear lanceolate to spathulate ovate and with a very strong midrib from which numerous fine veins pass out to the margin forming a network; the leaves are 3-40 cm long.

Q. 31. Name two gymnospermous plants producing balsam/resin.

Ans. *Pinus roxburghii, Canada balsam.*

Q. 32. What are the differences between pollen grains of *Gnetum* and *Ephedra*?

Ans. The pollen grain in *Ephedra* are round or elliptrical. The pollen dehisces by transverse or oblique slits. Pollen grain of *Gnetum* has a spiny thick exine and an intime. They are libarated by longitudinal dehiscence of the anthers.

Q. 33. What are the major differences between pycnoxylic and manoxylic wood?

Ans.

Pycnoxylic	*Manoxylic*
1. It is compact and has narrow medullary rays.	1. It is porous, soft, and more parenchymatous in nature, has wide medullary rays.
2. It is found in Coniferophyta.	2. It is found in Cycadophyte.

Q. 34. Why is *Ginkgo* called a `living fossil'?

Ans. The living species *Ginkgo biloba* originated in the Triassic period of the Mesozoic era and since then it has continued to live unchanged and undeltered by the innumerable environmental hazards. So it is called living fossil.

Q. 35. Mention the xerophytic characters of *Pinus* needle.

Ans. (i) The epidermis is highly cuticularised.

(ii) Have sunken stomata.

(iii) The guard cells of stomata cells are overlapped by subsidary cells.

Q. 36. What are the characteristic features of *Kaloxylon*?

Ans. *Kaloxylon* is the root genera of *Lyginopteris oldhamia*.

It is adventatious root that was probably proproot supported by the weak stem. This possesses radial vascular strand where stele is tetrach with exarch xylem.

Q. 37. What are the distinguishing features of *Crossotheca*?

Ans. In *Crossotheca* the fertile pinnules are present. Each pinnule has 6 or more peltate, bilocular microsporangia. It contains microspores which are 50-70μ in diameter. Spore proably germinate in the pollen chamber.

Q. 38. How many cotyledons are there in the embryos of *Pinus, Cycas* and *Ginkgo*?

Ans. *Pinus* – 3 to many cotyledons.

Cycas – 2 cotyledons.

Ginkgo – finally 2 or sometimes 3 cotyledons.

Q.39. Name an edible gymnospermous seed.

Ans. *Ginkgo biloba.*

Q. 40. Where do you find the true vessel in gymnosperm?

Ans. In *Gnetum* sp.

Q. 41. What is endophytic association?

Ans. Some members of blue green algae, possibly *Anabaena* sp., grow within the root cortex of *Cycas* sp. The association between algae and gymnospermous root is known as endophytic association.

Q. 42. Where do you find endosperm as a post-fertilisation structure in Gymosperm?

Ans. *Gnetum* sp.

Q. 43. Mention two gymnospermous features in *Gnetum*.

Ans. (i) Presence of prothallial cell in the male gametophyte.

(ii) Naked seeded condition.

Q. 44. Where do you find nuclear embryo sac in gymnosperm?

Ans. *Gnetum* sp.

Q. 45. What are the different organ genera of the reconstructed plant *Williamsonia sewardiana*?

Ans. Leaf — *Ptilophyllum sahnii*

Stem — *Bucklendia indica*

Flower — *Sturiella.*

Q. 46. Where in India does *Ephedra* occur naturally?

Ans. (i) *Ephedra foliata* in Punjab and Rajasthan.

(ii) *Ephedra intermedia* in Kashmir, Punjab, Himachal Pradesh.

Q. 47. Do you know any gymnosperm which yields timber?

Ans. *Ginkgo biloba, Pinus roxburghii.*

Q. 48. What is plicate mesophyll?

Ans. In the needle leaves of *Pinus* the mesophyll tissue contains several infoldings and due to this they are often called plicate mesophyll.

Q.49. Who proposed the class Progymnospermopsida?

Ans. Bech (1960) proposed the class Progymnospermopside.

Q. 50. Name a parasitic gymnosperm.

Ans. *Gnetum trinerve.*

Q. 51. Name an aquatic gymnosperm.

Ans. *Taxodium* sp.

Q. 52. What is *Lagenostoma*?

Ans. Lagenostoma is a beak-like outgrowth of nuceller tissue within the micropyle found in the seed of *Lygenopteris.*

Q. 53. Where do you find sclerotic nest?

Ans. In the pith region of the stem *Lygenopteris.*

Q. 54. In which genus is the largest ovule of the plant kingdom found?

Ans. *Cycas siamensis.*

Q. 55. What is heterogeneous ray?

Ans. Heterogeneous ray appear in between the bundles where extra amount of mechanical strength are necessary. Usually they are formed by the parenchyma when intermingled with trachides or wood fiber. E.g. *Pinus* sp.

Q. 56. What is monosulcate pollen?

Ans. When the pollen consist of a furrow-like aperture situated at right angles to the polar axis then this type of pollen is called monosulcate pollen.

9 PALEO-BOTANY

Q. 1. Why the class Progymnospermopsida was formed?

Ans. On obtaining *Archaeopteris* frond attached to *Callixylon* stem, a new class Progymnospermopsida was proposed.

Q. 2. What is the fossilisation process of coal-ball formation?

Ans. Infiltration process.

Q. 3. Name earliest eukaryote appeared?

Ans. *Grypania* (2100 million)

Q. 4. Which geological period is usually considered as the period of early land plants?

Ans. Quaternary period of Cenozoic era.

Q. 5. What is diagenesis?

Ans. The chemical and physical process influencing sediment before and during rock formation is called diagenesis. It is extremely important in fossilization process.

Q. 6. Who reconstructed *Williamsonia sewardiana*?

Ans. Prof. Birbal Sahni.

Q. 7. When did *Rhynia* appear on the earth?

Ans. Middle Devonian period.

Q. 8. Which type of fossil preserves internal structure?

Ans. Petrified fossil.

Q. 9. What is an "index fossil"?

Ans. Certain fossils are known to be restricted to a particular period of geologic history and therefore contribute a very important means of stratigraphy. Such fossils are known as `index fossils' or, `guide fossils'.

Q. 10. Mention the periods of the Palaeozoic era sequentially.

Ans. (i) Devonian,

(ii) Silurian,

(iii) Ordovician,

(iv) Cambrian.

Q. 11. What is the name of a Bennettitalean plant reconstructed by Prof. Birbal Sahni?

Ans. *Wiliamsonia sewardiana.*

Q. 12. In which geological horizon stromatolite is abundant?

Ans. In the precambrian period of protenozoic era, i.e. the age of Cyanophyceae.

Q. 13. What type of rock is coal?

Ans. Organo-sedimentary rock having 50% or more carbon.

Q. 14. Name a heterosporus species of *Archaeopteris.*

Ans. *Archaeopteris latifolia.*

Q. 15. State the geological age of occurence of the fossil plant *Lyginopteris oldhamia.*

Ans. Carboniferous.

Q. 16. Name the index megafossil of lower Gondwana.

Ans. Pteridosperm *Glossopteris.*

Q. 17. Wherefrom were *Williamsonia sewardiana* plant parts discovered?

Ans. Leaves → Amrapara, Brindabon of Rajmahal hill.

Stem → Amrapara of Rajmahal hill.

Fructification → Amrapara of Rajmahal hill.

Q. 18. What is parichnos?

Ans. The actual leaf scar is above the middle line of the rhombus. In the leaf scar the bundle scar mark is flanked by two scars and at a lower level by two more larger scars. These four are the parichnos scars.

Q. 19. What is a fossil?

Ans. Fossil is the evidence of pre-historical life. Fossil is the retic or trace of a former living thing preserved in rock.

Q. 20. What is coal-ball?

Ans. Coal-balls are irregularly rounded concrete masses, commonly made up of calcite (calcified) containing preserved fragments of coal-forming plants. Each coal-ball is of localised occurence and contains $CaCO_3$, $MgCO_3$ and sometimes FeS.

Q. 21. Explain the importance of petrification in paleobotanical studies.

Ans. In case of petrification both the external and internal structure of the plant is preserved in original form and in which every cell is preserved perfectly distended as if in life so that it helps the study of the plant thouroughly, about its period of occurence, habit and habitat.

Q. 22. Name the different parts of a reconstructed plant.

Ans. Leaves → *Prilophyllum cutchens.*

Stem → *Bucklandia indica.*

Fructification → *Williamsonia sewardiana.*

Q. 23. Short note on Interseminal scales in *Williamsonia.*

Ans. The interseminal scales are clavate, distal and covered by a thick cuticle. Vascular bundles seem to reach the tip.

Q. 24. What is chemical fossil?

Ans. This type of fossil is preserved chemically are not morphologically.

E.g. – *Porphyrin.*

Q. 25. Who instituted the class Progymnospermopsida?

Ans. Beck (1960, 1962, 1968).

Q. 26. Secondary growth in *Lepidodendron?*

Ans. In the secondary growth of *Lepidodendron* is present an outermost hard periderm formed early in the life of a plant by a phellogen in the outer cortex. Outer cortex with periderm is the main supporting tissue of the trunk.

Q. 27. What is paleopalynology?

Ans. The study about the fossil pollen and spores including structure, formation dispersal. Its preservation is called paleopalynology. This term was coined by Hyde and Williams (1945).

Q. 28. State the geological and geographical distribution of *Bennettitales.*

Ans. *Bennettitales* were found from Triassic to lower Cretaceous but were predominen in Jurassic period of Mesozoic era in South America, North America, India and several other places – mainly in the southern hemisphere.

Q. 29. Why *Ginkgo biloba* is called a `living fossil'?

Ans. *Ginkgo biloba,* a monotypic species, seems to have been saved from extinction fo foreseeable future for it is a plant that has been reintroduced and cultivated in many parts of the world where it once flourished many million of years. So, G *biloba* is called a living fossil.

Q. 30. Mention the geological age and place of occurence of *Williamsonia sewardiana* and name the different forms of genera of it.

Ans. Geologic age –Jurassic

Place – Rajmahal hills.

Form genera of Leaves – *Prilophyllum cutchens*

Form genera of Stem – *Bucklandia indica*

Form genera of Fructification – *Williamsonia sewardii.*

Q. 31. Enumerate the four major plants of paleozoic era.

Ans. Marine algae (Dominant), primitive land plants, primitive pteridophytes and lycopods.

Q. 32. Mention the geological age of occurence of Cordaitales naming its different genera.

Ans. Some middle devonian tree trunks have been ascribed to this order but they are widely known from the permocarboniferous; they became extinct in the Jurassic.

In family Pityaceae of this order, 3 genus are

1. *Callixylon,*
2. *Archaeopitys,*
3. *Pitys.*

In family poroxylaceae the genus in *Poroxylon.*

In family cordaitaceae it is considered that this family contains only *cordaites* genus but this genera have some form genera like –

Form genera of wood → *Cordaixylon, Mesoxylon*

Form genera of root → *Amyelon*

Form genera of strobili or inflorescence → *Cordaianthus.* etc.

Q. 33. Describe the stem anatomy of *Lyginopteris* (Palaeozoic pteridosperm).

Ans. Stem anatomy of Lyginopteris shows :

(i) Outer cortex with radial patches of sclerenchymatous cells.

(ii) Parenchymatous inner cortex.

(iii) Thick ring of secondary wood with rows of radially arranged multiseriate bordered pitted tracheids.

(iv) Inside the secondary wood are 5 primary mesarch xylem strand.

(v) In some young stem there present primary and secondary phloem and cambium which are absent in older stem.

(vi) Leaf traces originating from primary xylem.

(vii) Pith lain patches having some sclerenchymatous patches are called *Sclerotic nests.*

Q. 34. Comment on *Rhynia's* geologic distribution.

Ans. It is known only from a single middle Devonian deposition.

10 MORPHOLOGY

Q. 1. What is phyllotaxy?

Ans. The arrangement of leaves on the stem is not random, ways – there is a particular manner. The arrangment of leaves on the stem is known as phyllotaxy.

Q. 2. What is ligule?

Ans. In many plants of the family Graminae there is no petiole. Some peculiar outgrowths grow at the junction of the sheathing leaf-base and the lamina. This is tongue like membranous structure called ligule.

Q. 3. What is reticulate venation?

Ans. The strong main vein gives rise to finer branches which again branch repeteadly running into one another and forming a reticulum or network all over the leaf called reticulate venation.

Q. 4. What is parallel venation?

Ans. When the main veins tend to run parallel to one another and are connected laterally by transverse veinlets which are unbranched and parallel to one another it is called parallel venation.

Q. 5. What do you mean by convergent?

Ans. When the prominent vein, after spreading out from the base, again converges towards the tip of the lamina it is called convergent.

Q. 6. What do you mean by divergent?

Ans. When the costa spread out farther and farther apart as in the Cucurbita it is known as divergent.

Q. 7. What is hydathode?

Ans. Hydathode occurs on leaves and are associated with the marginal cells of the leaves. Through the hydathode the exudation of the water under condition of low transpiration and high root pressure take place. E.g. *Colocassia* sp.

Q. 8. Define hypanthium.

Ans. The cup-like receptacle derived usually from the fusion of floral envelops androccium on which the calyx, corolla and stamens are born.

Q. 9. Define regma and cite an example.

Ans. Regma is a type of dry fruit which develops from a superior tri- or polycarpelary ovaries. The fruit splits up into many parts called cocci. Each part contains one or two seeds, which are liberated on decaying of the dry pericarp. E.g. *Ricinus communis.*

Q. 10. What is cymba? Give an example.

Ans. Cymba is a woody durable, boat-like spath or spath valve that encloses the inflorescence, open and persist. E.g. Cymba found in many palms.

Q. 11. Define heterostyle and give one example.

Ans. In certain plants there are flowers of two or three different form with anther and stigma at different levels. This take forms place by styles of different length. This is called heterostyle. E.g. *Primula sinensis.*

Q. 12. What is hypanthodium? Cite example.

Ans. The inflorescence hypanthodium is composed of three fused primary axes or rachis of closely placed cymes, as a result of such fusion a cup-shaped fleshy concave receptacle almost or completely closed structure with a small opening at the apex result. The unisexual flowers are arranged all over the inner surface of the receptacle in cymose group. E.g. *Ficus* sp.

Q. 13. What is gynostagium? Give an example.

Ans. In some family Asclepiadaceae, Orchedaceae, the stamens and carples unite completely forming columnar structure which is called gynostagium.

E.g. *Calotropis procera* (Asclepiadaceae)

Q. 14. Define `achene'?

Ans. It is a indehescence dry fruit. It is one-seeded and develops from a superior ovary. Here pericarp and the seed coat are free from one another. E.g. Achene may be seen in *Mirabilis jalapa.*

Q. 15. What is corymb?

Ans. Corymb is an inflorescence. A corymb is a typical raceme with unequal flower stalk, i.e. pedicels, so that the basal flower stands on the same level as upper one. E.g. *Cassia* sp.

Q. 16. Distinguish between syngenesious and synandrous stamens.

Ans. In syngenesious condition, in an androecium the filament are free but anthers are united to form a tubular structure round the style. E.g. Compositae family. When stamens are united by filaments as well as anther to form a compact body it is known as synnandrous. E.g. In the family Cucurbitaceae.

Q. 17. What are pseudocarp? Cite one example.

Ans. Sometime the others part of a flower also take part in the formation of fruit. They are together with ovary. In this case the true part of the fruit remains insignificant and tiny. Other parts such as receptacle calyx etc. entering into its composition form the major part of the fruit.

Q. 18. Define `Coma' cite example.

Ans. Coma is a tuft of hair developed as a crown on the seed of most plants of the family Apocyanaceae and Asclepiadaceae. E.g. *Calotropis procera.*

Q. 19. What is personate corolla?

Ans. Personate corolla is a kind of bilabiate in which the throat of corolla is closed by a projection of lower lip known as palate. E.g. *Antirrhinum* sp.

Q. 20. Name a plant having dimorphic flower.

Ans. *Primula sinensis.*

Q. 21. What is aril?

Ans. After fertilisation an outgrowth looking like a third integument is sometimes seen to develop from the funicular end. E.g. *litchi chinensis.*

Q. 22. What is retinaculum? In which family is it found?

Ans. Seed of the Acanthaceae are often scaled on hook-like outgrowths known as retinaculum or jaculator. It is found in family Acanthaceae (*Ruallia* sp).

Q. 23. What are cremocarp?

Ans. Cremocarp is the two-seeded fruit which develops from the interior bicarpallary ovary of Apiaceae. Here the two mericarps split along the carpophore to which they are attached.

Q. 24. What is apocarpous pistil?

Ans. When the carples are completely free from one another they they are known as apocarpous pistils. E.g. Found in family Magnoliaceae.

Q. 25. What is pepo? Where is it found?

Ans. Pepo is a simple, fleshy many-seeded fruit which develops from a polycarpelary superior ovary with axile placentation. Here the epicarp and mesocarp are fused together forming the skin of the fruit. E.g. *Cucurbita* sp.

Q. 26. What is leaf Mosaic?

Ans. Plants growing in shady place, where exposure to light is minimum, have a wonderful display of large and small leaves and, as a result, of continuous flat green surface is formed. E.g. *Acalypha* sp.

Q. 27. What is a polygamous plant?

Ans. If staminate, pistillate and hermaphrodite flowers develop on the the same plant, this plant is known as Polygamous plant. E.g. *Mangifera indica.*

Q. 28. What do you mean by lomentum?

Ans. This is a variation of legume. This fruit is constricted between the seed and usually breaks up into bits containing one or more seed(s). This is seen in the family leguminosae. E.g. *Acacia* sp. (Mimoseae).

Q. 29. Define heterostyle.

Ans. In certain plants there are flowers of two or three different forms with anthers and stigma at different levels. When this takes place by different levels of style it is called heterostyle. E.g. *Primula sinensis.*

Q. 30. What is drupe?

Ans. This is normally a one-seeded fruit with a skinny epicarp, a fibrous and usually juicy mesocarp and a stony hard endocarp. E.g. *Mangifera indica.*

Q. 31. What is corona?

Ans. When the appendages of the corolla are in the form of a scale, hair-like outgrowths of various types are situated at the base of petals or in-between clow and limb. This is the Corona. E.g. *Passiflora* sp.

Q. 32. Define sorosis with example.

Ans. It is a multiple fruit. Here the rachis and all the floral parts of a spadix type of inflorescence fuse together forming a composite fruit. E.g. *Artocarpus heterophyllus.*

Q. 33. What is phyllodes and cladodes?

Ans. Phyllodes are green modified flattened or winged petiole. Here leaflets fall off very early or they do not develop at all.

E.g. *Acacia auriculoformis.*

Cladode is a metamorphosed aerial stem. Here each branch consist of one internode arising at the axil of minute scale leaves.

E.g. *Asparagus racemesus.*

Q. 34. What is stylopodium?

Ans. When the base of the style is swollen it is called stylopodium. It is found in the family Umbelluferae. E.g. *Apium* sp.

Q. 35. What is cypsela?

Ans. It is a type of indehesecent dry fruit. It develop from the inferior ovary and has a pericarp free from the seed coat. E.g. *Helianthus annus* (Compositae).

Q. 36. What is syncarpous?

Ans. When the carples are fused with each other then the condition is known as syncarpous. E.g. Family Cruciferae.

Q. 37. What is aestivation?

Ans. The arrangement of sepals or petals in a flower is known as aestivation.

Q. 38. What is gynobasic?

Ans. When the style appear to arise from the central base of the ovary it is called gynobasic.

Q. 39. What do you mean by anatropous and orthotropus ovule?

Ans. When the ovule is reversed type – the commonest – and may be seen in pea, bean, castor etc.

When the ovule remains straight in Polygonoceae, Piperaceae it is called orthotropous ovule.

Q. 40. What is ornithophilous flower?

Ans. Bird-pollinated flower is known as ornithophilous flower.

E.g. *Bignonia capreolata.*

11 ANATOMY

Q. 1. What is stone cell?

Ans. Stone cell is a type of sclerides. They are paranchymatous cell and are usually isodiametric. They are found in fleshy fruits. E.g. *Psidium* sp, *Pyrus* sp. etc.

Q. 2. What is leptocentric vascular bundle?

Ans. When in a vascular bundle the phloem is surrounded by the xylem it is called leptocentric vascular bundle. E.g. Monocot stem of *Dracaena.*

Q. 3. What are intraxylary phloem?

Ans. When the phloem is present on the inner side of the xylem it is called intraxylary phloem. E.g. In *Leptadenia.*

Q. 4. What is cortical bundle?

Ans. When vascular bundle occur in the cortex region, they are called cortical bundle. E.g. Cortical bundle found in families like Cactaceae, Oleaceae.

Q. 5. What do you mean by medullary rays?

Ans. Medullary rays is the tissue region lying in between vascular bundle and thus occupying the inter fascicular region. Cells of medullary rays are elongated, they often establish the connecting link between the cell of cortex and pith.

Q. 6. What is casparian strips?

Ans. The radial and inner wall of the endodermal cell are thick and the outer wall are thin. Sometimes the thickening of the radial wall is in the form of a strip called casparin strip. In transverse section it looks like large dots.

Q. 7. What is tonoplast?

Ans. A vacuale is surrounded by a single layered enveloped called tonoplast.

Q. 8. What is phelloderm?

Ans. Phelloderm is the tissue formed by the phellogen in the opposite direction from the cork. Resembles the cortical parenchyma. It is a part of periderm. The cells are both photosynthetic and storehouse of carbohydrate.

Q. 9. What do you mean by dictyostele?

Ans. In the more advanced siphonostele of *Pteropsida* the successive leaf gaps may overlap each other. Birebner (1902) called siphonostele with leaf gaps dictyostele.

Q. 10. What do you mean by transfusion tissue?

Ans. In gymnosperm leaves, a tissue surrounding or associated with the vascular bundle, and composed of non-living tracheides are called transfusion tissue. E.g. They are present in *Cycas* leaf.

Q. 11. Define rhytidome.

Ans. Rhytidome is the dead part of the bark composed of tissue layers rotated from the periderm; it is also called outer bark.

Q. 12. Define tylosis.

Ans. Tylosis is balloon-like protuberance or outgrowth of part of cell wall of axial and ray paranchyma cell projecting inside the xylem vessel through pit cavities. They store starch.

Q. 13. What is promeristem?

Ans. Promeristem is the region of new growth in plant body where the foundation of new organ or inner part is initiated. It is composed of all young thin-walled cells which are alike.

Q. 14. What is bordered pits?

Ans. Bordered pit are more complex and more variable in structure than simple pits. In this type pits are provided with a rim or border, the secondary wall develops over the pit cavity i.e. pit chamber to form an over-arching roof with a narrow pore in the center.

Q. 15. What do you mean by medullary bundle?

Ans. When the vascular bundles are scattered in the pith these bundles are referred to as meduallry bundle. They are usually amphivasal. E.g. families like Oleceae, Cactaceae.

Q. 16. What is plasmodesmata?

Ans. Plasmodesmata is a strand of cytoplasm passing through the cell wall and usually joining the protoplast of the two adjacent cells. They are concerned with the translocation of the food.

Q. 17. What do you mean by trichome?

Ans. All unicellular and multicellular appendages of the epidermis are designated by the term trichome. E.g. Hair, bladders, root hair etc.

Q. 18. Define heartwood.

Ans. The central region of the old tree, which was formed earlier and filled up with tannins, resin, gum and other substances which make it hard and durable, is called heartwood. It look black due to presence of various kinds of substance.

Q. 19. Define vestured pits.

Ans. Vestured pits are special type of pits named by Baily (1913). In some dicotyledon thin, minute, simple or branch sculptures are present on the secondary wall of the pit chamber or along the edges of the pit aperture.

Q. 20. What is paracytic stomata?

Ans. In paracytic type of stomata, the stomata remain surrounded by two subsidiary or accessory cells which are parallel to the long axis of the pore and guard cells. E.g. This type of stomata found in the family Rubiaceae.

Q. 21. What is torus?

Ans. In borderd pit, the central thickened part of the pit membrane consisting of middle lamella and two primary walls is known as torus.

Q. 22. Define lenticel.

Ans. Lenticel are restricted area of relatively loosely arranged suberise or non-suberised cells in the periderm. In a lenticel the phellogen is more active than elsewhere and producing tissue have much intercellular space.

Q. 23. What do you mean by amphivasal and amphicribal vascular bundle?

Ans. In amphivasal vascular bundle phloem is surrounded by xylem. Some monocot stems like *Dracaena* have this kind of vascular bundle. They are always closed. In amphicribal vascular bundle the xylem is surrounded by phloem. They are found in stem of pteridophyte like *Lycopodium, Selaginella.*

Q. 24. What is leaf gap?

Ans. In the nodal region of a leaf where a leaf trace bends away from the centre of the stem towards the leaf branch a parenchymatous region continuous with pith occurs in the vascular cylinder of the stem. The region which appear as a interfascicular region in transverse section is known as leaf gap.

Q. 25. What are complementary cell?

Ans. Complementary cells are non-suberised cells loosely arranged in cork cells and forming air passage. They are protective in function.

Q. 26. What is storied cork?

Ans. Storied cork is the protective tissue formed in the stem of perennial monocots. In thickned stem of monocot a special type of protective tissue develops. This tissue is formed by the activity of storied meristem. The meristem is also suberised and is called storied cork.

Q. 27. Define bicollateral vascular bundle; cite example.

Ans. In this type there are two strands of phloem and a central xylem. The two pholem strands are separated from central xylem strand by two strips of cambium. This always remains open due to the presene of cambium. E.g. Stem of member Cucurbitaceae.

Q. 28. What is phellem? State its function.

Ans. It is the protective tissue composed of non-living cells with suberised walls. It replace the epidermis in one year, older stem and roots of many plants are formed by the phellogen. Function – It is protective.

Q. 29. What do you mean by aleurone layer?

Ans. This is a storage layer made up of aleurone present in endosperm.

Q. 30. What is periderm?

Ans. Secondary protective tissue derived from the phellogen and replacing the epidermis, typically in stems and roots. Consisting of phellem, phellogen and phelloderm.

Q. 31. What is actinostele?

Ans. It is a type of protostele; star shaped xylem is surrounded by phloem. E.g. stem stele of *Lycopedium serratum*.

Q. 32. What do you mean by lateral meristem?

Ans. Lateral meristem are situated laterally i.e. parallel to the circumerence of the organ in which they occur. This meristem is known as lateral meristerm; it is always composed of a single layer of rectangular cell.

Q. 33. Name a root having anomalous structure.

Ans. Storage root of *Beta vulgaris* (Chenopodiaceae).

Q. 34. Name the principle of distribution of mechanical tissue in the plant body.

Ans. (i) Inflexibility.

(ii) Inextensibility.

(iii) Incompressibility

(iv) Shearing stresses.

Q. 35. What do you mean by resin duct?

Ans. Substance like resin and oil are copiously secreted in gymnosperm in general and also in many angiosperms. The material are secreted and conducted through the special glands which are called ducts. The resin ducts are tube-like bodies having a lining of small parenchyma cell with dense protoplast.

Q. 36. Define annual ring.

Ans. The activity of combium is often periodical, and, as a result, distinct growth layers are formed in xylem. In most cases the periodicity in the activity of the cambium is seasonal. So the secondry xylem formed in a year constitutes the growth ring known as annual ring.

Q. 37. Define sap wood and heart wood.

Ans. After formation of considerable quantity of secondry xylem two types of wood appear in the stem. The outer portion consists of recently formed xylem called sap wood and the centrally located portion, which are formed earlier, are called heart wood.

Q. 38. What is caryophyllaceous stomata?

Ans. When the stomata remain enclosed by a pair of subsidiary cells whose common wall is at right angle to the guard cell it is known as caryophyllaceous stomata—found in the family Acanthaceae, Caryophyllaceae.

Q. 39. What is scleride?

Ans. Scleride are non-parenchymatous cells, usually isodiametric and irregular in shape. Specially occur in fruits and seed coat.

Q. 40. What is trabaculae?

Ans. In the secondary wood of conifers rod-like or bar-like sculpturing are common in the lumen of the tracheids extending across from one tangential wall to another – these are called trabaculae.

Q. 41. Define crassulae.

Ans. These are characteristic thickening of the middle lamella and primary walls which are visible through the secondary wall in stained preparation. They are called crasulae.

Q. 42. What is raphides?

Ans. Raphides are calcium oxalate crystals present particularly in higher plants. They are needle like and found abundant in storage organs.

12 TAXONOMY

Q. 1. What is biosystematic phase?

Ans. Biosystematic phase is the part of classification showing phylogeny and relationship among the group of plants by cytological or embroylogical evidence of plants.

Q. 2. What is basionym?

Ans. The combination which is based on an earliest legitimate name, the earliest legitimate name is called basionym.

Q. 3. Name the type genus of Umbelliferae.

Ans. The type genus of Umbelliferae (Apiaceae) is *Apium* sp.

Q. 4. What is Adansonian taxonomy?

Ans. Michel Adanson, a French botanist, included the rejection of all the artificial classification for natural and the description of taxa more or less equivalent to modern orders and families. The taxonomy which is based on Adanson principle is called Adansonian taxonomy.

Q. 5. Write the name of the plant having labellum in the flower.

Ans. *Vanda tessellata* is a plant which has labellum in the flower.

Q. 6. Name a plant having unistaminate flower.

Ans. *Vanda roxburgii* (orchedaceae) having unisataminate flower.

Q. 7. Name a plant having free central placentation.

Ans. *Primula* sp (Primeelaceae).

Q. 8. What is tautotaxonomy?

Ans. When the specific name repeat alternatively the generic name then it is known as tautotaxonomy. E.g. *Malus malus*.

Q. 9. Name a medicinal plant of family Rubiaceae.

Ans. *Cinchona succirubra* is the medicinal plant of the family Rubiaceae which is the source of quinine.

Q. 10. Name a plant of Scrophulariaceae with regular flower.

Ans. *Scoparia dulsis.*

Q. 11. What is the full form of ICBN?

Ans. International Code of Botanical Nomenclature.

Q. 12. Name a plant of umbelliferal having edible roots.

Ans. *Daucus carota* (gajor) – the root of the plant used as vegetable.

Q. 13. Name a monocot family having exalbuminous seeds.

Ans. Musaceae have exalbuminous seed.

Q. 14. Write two advanced reproductive features of Asteraceae.

Ans. (i) The syngenesious protandrous androcium together with epigyny are no doubt advanced in character from the evolutionary point of view.

(ii) The ovary is inferior with a single basal ovule.

(iii) The inflorescence is capitulum.

Q. 15. Cite the example of an amenophilous plant.

Ans. *Oryza sativa* (Poaceae).

Q. 16. What is omega taxonomy?

Ans. Classification showing phylogeny and relationship among group of plants by cytological, embryological evidence of plant is called omega taxonomy.

Q. 17. Name a plant having dimorphic anther.

Ans. *Primula sinensis.*

Q. 18. Name a plant family characterised by basal placentation.

Ans. The family Compositae.

Q. 19. What is acronym?

Ans. Standard abbreviation of Botanical Names are called acronyms.

Q. 20. In which family is `stylopodium' found?

Ans. In he family Umbelliferae stylopodium is found.

Q. 21. What is meant by NPC system?

Ans. This is a system of classification of pollen grain on the basis of their number (N), position (P) and characters (C).

Q. 22. Mention the most salient reproductive characters of Cucurbitaceae.

Ans. (i) Stamens forming synandrium.

(ii) Flowers are unisexual.

Q. 23. What do you mean by α-taxonomy?

Ans. When the classification is based on the external morphological characteristics then such taxonomy is known as α-toxonomy.

Q. 24. What are botanical gardens?

Ans. Botanical gardens are essentially collections of living plants maintained for both pure and applied studies.

Q. 25. Name the largest Botanical Garden in India and mention its location.

Ans. The largest Botanical Garden in India is Indian Botanical Garden (IBG) near Kolkata at Howrah (Sibpur).

Q. 26. What is the formula of the family Malvaceae?

Ans. $\oplus$ ⚥ Epi $K_{(5)}\ C_5\ A_{\alpha}\ G_{\underline{(5-\alpha)}}$

Q. 27. What do you mean by `key'?

Ans. Key is an artificial analytic device or arrangement whereby a choice is provided between two contradictory propositions – results in applying one and rejecting the other.

Q. 28. What is systemic botany?

Ans. Systemic Botany is the science which deals with the classification and naming of the plants.

Q. 29. Name a monocot family having dicot features.

Ans. Aroceae (*Colocasia esculenta*) having reticulate venation.

Q. 30. Mention the speciality of calyx in the flower of Compositae.

Ans. In Compositae the calyx is reduced to form hairy structures called pappus.

Q. 31. What is the alternative name of the family Palmae?

Ans. Arecaceae.

Q. 32. Name a climbing plant from the family Apocyanaceae.

Ans. *Aganosma dichotoma.*

Q. 33. Name a dicot plant having pollinia.

Ans. *Calotropis procera.*

Q. 34. Name an oil-yielding monocot plant.

Ans. *Oryza sativa.*

Q. 35. What do you mean by monotypic family?

Ans. The family which contains single genus is called monotypic family. E.g. In monocot the family Canaceae presents only one genus *Canna.*

Q. 36. Write two advanced features of the family Orchidaceae.

Ans. (i) The zygomorphic flowers together with the various structural modification.

(ii) Presence of labellum is one of the highest morphological adaptation for cross-pollination.

Q. 37. Name the largest Botanical Garden of the world.

Ans. Royal Botanical Garden, Kew, London.

Q. 38. What is taxon?

Ans. A taxonomic group of any rank is called a taxon (plural : taxa); such as a phylum, a family or a species. e.g. Orchidaceae (a famila), *Calotropis* (a generic taxon).

Q. 39. Why the Linnaean system of classification treated as an artificial system?

Ans. In an artificial system of classification organisms are classified by means of one or a few superficial characteristics. Linnaeans classified the plants on the basis of stamen's characteristics which are of superficial character, so it is an artificial system.

Q. 40. Cite one vegetative and one reproductive characteristics of the member Malvaceae which is identifying.

Ans. Veg → Stipules are free lateral, epicalyx present.

Rep → Stamens monodelphous, with a distinct staminal column.

Q. 41. Name a latex-yielding plant of the family Euphorbiaceae.

Ans. *Hevea brasiliensis.*

Q. 42. What is the alternative name of the family Leguminosae?

Ans. Fabaceae.

Q. 43. Write two primitive features found in the members of the family Magnoliaceae.

Ans. (i) Stamens are many, free and spirally arranged.

(ii) Leaves are simple, alternate and stipulate.

Q. 44. What is the fruit type of the family Labiatae?

Ans. Fruits of Labiatae consist of 4 one-seeded nutlets.

Q. 45. What is "natural taxon"?

Ans. When a plant is fully expended reproducing, migrating and expanding in its area of occurence it is called `natural taxon'.

Q. 46. What do you mean by holotype?

Ans. A particular illustration is designated by the author of the species to represent the type of a species called holotype.

Q. 47. What is the full form of IAPT.

Ans. International Association of Plant Taxonomy.

Q. 48. What is nomen nudum?

Ans. The names with no accompanying description are called nomen nudum.

Q. 49. Name some economically important plants of Amaranthaceae.

Ans. (i) Medicinal plant : *Acrva aspera, A. lentana.*

(ii) Ornamental : *Celosia cristata.*

(iii) Leaf as vegetables : *Amaranthus tricolour.*

Q. 50. What do you mean by isotype?

Ans. A specimen which is duplicate of holotype, collected from the same place, at same time and the same person is known as isotype.

Q. 51. What is numerical taxonomy?

Ans. The analysis of various types of taxonomic date by mathematically computerised method is called numerical taxonomy.

Q. 52. What is chemotaxonomy?

Ans. The approach of taxonomy in which chemical features of plant are used in developing classification or in solving taxonomic problems is called chemotaxonomy.

Q. 53. What is syntype?

Ans. Any one of the two or more specimens cited by the author when no holotype was designated.

Q. 54. What is neotype?

Ans. When all the material missing neotype are selected from other material.

Q. 55. What is effective publication?

Ans. A publication become effective by distribution in printed forms through changes to give to general public or at least the botanical institutes.

Q. 56. What is parallelism?

Ans. It refers to parallel evolution when independent evolutionary changes occur in groups with a common ancestry.

Q. 57. Name the year of publication of the latest version of ICBN.

Ans. In 1999.

Q. 58. Name a book along with its author which is commonly consulted for identification of angiosperm.

Ans. "Bengal plants" the author is David Prain.

Q. 59. Name a plant with coma.

Ans. *Calotropis procera* (Asclepiadoceae).

Q. 60. Mention the most salient reproductive character of Euphorbiaceae.

Ans. The gynocium is 3 carpled, ovary superior, 3 chambered each containing one or 2 pendulous ovules with ventral raphe.

Q. 61. Distinguish between monophyly and polyphyly.

Ans. The derivation of a taxon through one or more lineages from one immediately ancestral taxon of the same on lower rank is called monophyly.

The derivation from more than one ancestors is called polyphyly.

Q. 62. Mention one vegetative and one reproductive character for identification of the family Rubiaceae.

Ans. Vegetative – Stipules are inter- and intra-petioler.

Reproductive – Axile placentation.

Q. 63. Name a vesselless angiosperm.

Ans. *Drymis* sp (Winteraceae).

Q. 64. Mention one vegetative and one reproductive character for identification of the family mulvaceae.

Ans. Vegetative – Stipules are free lateral, epicalyx present.

Reproductive – Stamens are monodelphous, with a distinct staminal column.

Q. 65. Write the floral formula of the family Rubiaceae.

Ans. $\oplus$ ⚥ $K_{(4\text{-}5)}\ C_{(4\text{-}5)}\ A_{4\text{-}5}\ G_{(2)}$

Q. 66. What type of fruit is found in the family Umbelliferae?

Ans. Cremo carps or nut-like.

MONOCOTYLEDONS

Name of the family	Type genus	Floral formula	Floral diagram	Economic importance
1. Gramineae or Poaceae	*Oryza*	% ⚥ ♂ or ♀ P_0 or $_2$ lodicules A_3 or $_6$ $G_{\underline{(1)}}$	 *Oryza sativa*	Gramineae is a family of great importance from the economic point of view. Plants of this family provides food for man, fodder for domestic animal, used as ornamental plants for our lawns and gardens. (i) *Oryza sativa* : Used as a common staple food by millions of people throughout the world, rice is also used as a livestock & for manufacturing paper, starch, alcohol etc.; its husk is used as fuel and its brain is the source of fatty oil. (ii) *Triticum aestivum* : Used as staple food for man. Its straw is used as stock feed for cattles, for packing goods, manufacturing paper.
2. Palmae	*Areca*	**Male —** ⊕ ♂ P_{3+3} A_{3+3} **Female —** ⊕ ♀ P_{3+3} $G_{\underline{(3)}}$ or $_{\underline{3}}$	 Male flr. of *Areca catechu*	This family is supposed to be 2nd in importance to poaceae. Difft. plant parts of difft. plant of this family are used in many ways. The stem is used as pillars, as fuel; leaves are used for making mats, fans, basket etc.; toddy, obtained from inflorescence, provides an intoxicating alcoholic drink. (i) *Areca catechu* : *(Supari or, betel-nut)* : The nuts are chewed along with betel leaves (pan), the husks are used for making

Name of the family	Type genus	Floral formula	Floral diagram	Economic importance
			 Female flr. of *Areca catechu*	hard board and brown wrapping paper, the trunk is used as water pipe, the leaves are used for thatching, nuts are used for making fancy articles. (ii) *Cocos nucifera* (Coconut or, Nariyal) : An edible oil, obtained from the nut, is used in making soap, candle, confectionary. The fibrous rind of the nut yields the coir, which is used for making carpets & ropes. The endosperm yields oil, used as hair oil. The inflorescence in young condition, fermented, gives alcoholic drinks. The root is astringent & used in uterine diseases.
3. Musaceae	*Musa*	% ⚥ P_{3+3} $A_{6\ or\ 5}$ $G_{\overline{(3)}}$	 *Musa paradisiaca*	Difft. plants of thin family have difft. importance from the economic point of view. (i) *Musa paradisiaca* : It is one of the most famous tropical fruits, eaten throughout the world. Green bananas are used as vegetable and dried banana powder is a good baby food used in the manufacture of biscuits, chocolate etc. Starch is obtained from stem. Fibre is obtained from the sheathing leaf base. (ii) *Strelitzia reginae* : It is cultivated as ornamental garden plants for their showy leaves & brilliantly coloured flowers.

Name of the family	Type genus	Floral formula	Floral diagram	Economic importance
4. Cannaceae	*Canna*	% ⚥ $P_{3+3}\ A_6\ G_{\overline{(3)}}$		The plants of this family are mainly important economically for their ornamental value. (i) *Canna indica* : It is cultivated as ornamental garden plants. (ii) *Canna edulis* : From the rhizome of this plant a kind of starch is prepared which is used as foodstuff.
5. Liliaceae	*Lilium*	⊕ ⚥ $P_{(3+3)}\ A_{3+3}\ G_{\underline{(3)}}$	 *Allium cepa*	This family is of little economic value. Some are grown as ornamentals while others possess medicinal properties. (i) *Allium cepa* (Onion) : The underground disc-like small stems (bulbs) of this plant yield foodstuffs, used as vegetables & condiments. (ii) *Aloe barbadensis;* Syn. *A. vera* (Ghikanvar) : This is a succulent shrub. The leaves yield a resinous drug, which is used mainly as purgative.
6. Orchidaceae	*Orchis*	% ⚥ $P_{3+3}\ A_{1\ or\ 2}\ G_{\overline{(3)}}$	 *Vanda roxburghii*	The family is of little economic value. Most of the plants are grown for their beautiful flowers. Some possess medicinal properties and some yield scent. (i) *Orchis latifolia* : This is a perennial herb found in Kashmir. The tuberous roots are edible. (ii) *Vanda roxburghii* : Epiphytic herbs found on mango trees. The roots are used in rheumatism.

DYCOTYLEDONS

Name of the family	Type genus	Floral formula	Floral diagram	Economic importance
1. Magnoli-aceae	*Magnolia*	$\oplus$ ⚥ $P_{\alpha} A_{\alpha} \underline{G_{\alpha}}$ or, $K_{\alpha} C_{\alpha} A_{\alpha} \underline{G_{\alpha}}$	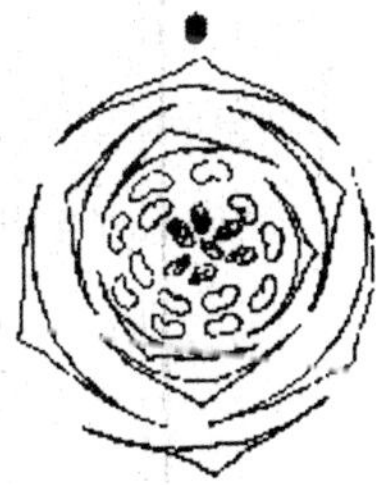 *M. campbelli*	The magnoliaceae are important as common ornamentals and a source of wood. (i) *Magnolia* sp : This plant is cultivated in gardens. *M stellata* are known for their sweet, scented, beautiful flowers. Wood of *M. acuminata* is used in making toys, tea boxes and quality furniture. (ii) *Michelia champaca* : It is a common garden ornamental plant, and source of a volatile oil used in perfumery and cosmetics. Its flowers and fruits are used to cure kidney troubles and gonorrhoea.
2. Ranuncu-laceae	*Ranuncu*lus	*R. muricathus* $\oplus$ ⚥ $K_5\, C_5\, A_{\alpha}\, \underline{G_{\alpha}}$	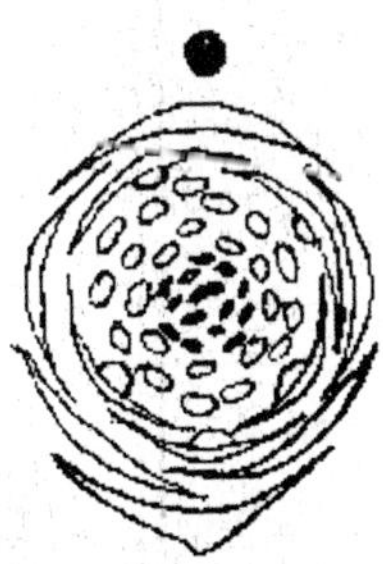 *R. sceleratus*	The family is of little economic value. Many plants are grown in gardens as ornamentals, some are of medicinal importance. (i) *Ranuculus aquatilis* : It is used medicinally in the treatment of rheumatic pain & asthma (ii) *Nigella sativa (kala jira)* : The seeds are used as condiments and speices. The seeds mixed with sessamum oil are used externally in eruption of skin & for scorpion sting.

Name of the family	Type genus	Floral formula	Floral diagram	Economic importance
3. Brassicaceae	*Brassica*	*B. campestris* $\oplus$ ⚥ $K_{2+2}\, C_4\, A_{2+4}\, \underline{G_{(2)}}$	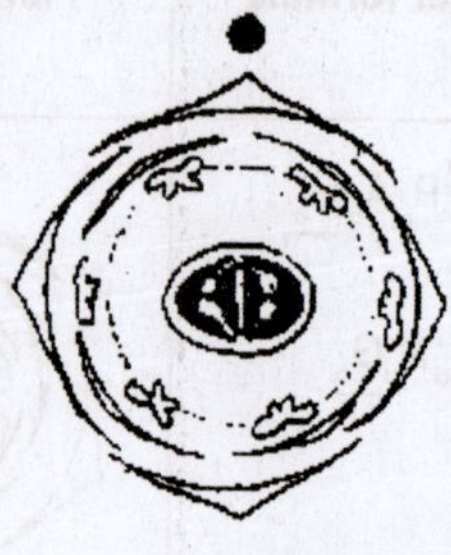 *B. nigra*	The family is fairly important from the economic point of view. Most of the plants contain sulphur compounds. The seeds of many plants yield vegetable oil of multipurpose use. (i) *Brassica compestris* var. *sarson* : It is an oilseed crop. The oil is used for cooking & burning purposes & the oil cake as a cattle feed. The tender leaves & shoots are used as vegetables. (ii) *Raphanus sativus* (Radish) : The roots, leaves are used as vegetables.
4. Malvaceae	*Malva*	*M. peniflora* $\oplus$ ⚥ Epi $K_3\, K_{(5)}$ $\overline{C_5\, A_{(\alpha)}}\, \underline{G_{(\alpha)}}$	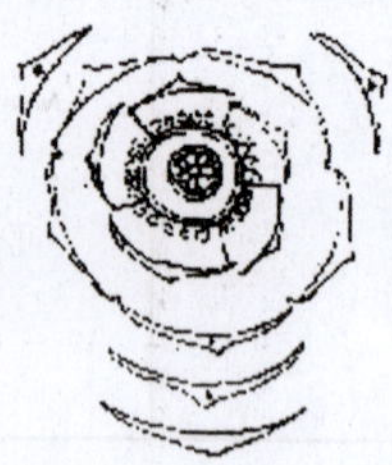 *Malva parviflora*	The family is of great economic value as it contains. Valuable fibre yielding plants but some plants have medicinal value. (i) *Abutilon indicum* : It is cultivated for its stem fibre, which is used for making ropes. (ii) *Abelmoschus esculentus* (Bhindi) : The fruits are used as vegetable & stalks are used in making fibres.
5. Cucurbitaceae	*Cucurbita*	Male flower — $\oplus$ ♂ $K_{(5)}\, C_{(5)}\, A_{(5)}$ Female flower — $\oplus$ ♀ $K_{(5)}\, C_{(5)}\, \overline{G}_{(3)}$	 *Male flower of C. pepo* 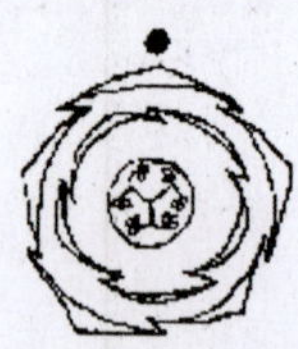 Female flower of *C. pepo*	Economically this family is fairly important. They possess edible fruits. The flrs, fruits are also used as vegetables, some plants have medicinal value. (i) *Cucurbita maxima* : This is grown for its large edible fruits, which are used as vegetables. (i) *Cucumis sativus* (Khira) : The fruits are edible and used in salad preparation.

Name of the family	Type genus	Floral formula	Floral diagram	Economic importance
6. Umbelliferae or Apiaceae	*Apium*	$\oplus$ ⚥ $K_{(5)} C_5 A_5 \overline{G}_{(2)}$	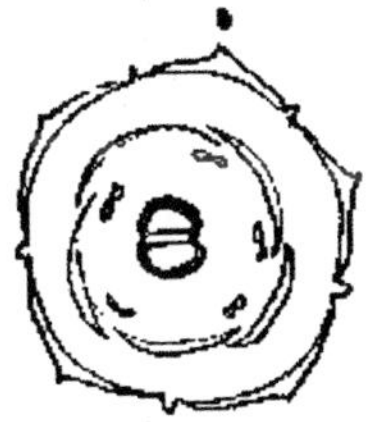	The family is fairly important from its economic point of view. The fruits of some plants are used as condiments whereas the other plants are of medicinal value. (i) *Foeniculum vulgare (saunf)* : The fruits of this plant are used as spice and also used medicinally as carminative and stimulant. (ii) *Centella asiatica (Brahmi)* : Used as an antidote against cholera and also to cure madness.
7. Euphorbiaceae	*Euphorbia*	*Euphorbia* sp. Male — $\oplus$ ♂ $K_0 C_0 A_1$ Female — $\oplus$ $K_0 C_0 \underline{G_{(3)}}$	 *Euphorbia pulcherrima*	The family consists of the plants of great economic value : (i) *Euphorbia hirta* : The plant is of medicinal value and used in many diseases of children and adults. (ii) *Ricinus communis (castor-oil plants)* : The seeds of this small tree are the source of castor-oil, which is mainly used as a lubricant and as a purgative. It is also used for textile soap, typewriter inks, perfume, varnishes & paints. The seed cake is used as fertiliser.
8. Acanthaceae	*Acanthus*	% ⚥ $K_{(4-5)} \overline{C_{(4-5)} A}_{4 \text{ or } 2}$ $\underline{G_{(2)}}$		The family is of little economic value. Some are used as hedge plants and some are ornamental and some plants have medicinal properties. (i) *Adhatoda vasica* : An Ayurvedic drug is obtained from its leaves which is used in cough, chronic bronchitis,

Name of the family	Type genus	Floral formula	Floral diagram	Economic importance
				rheumatism. The twig and leaves are used as green manure. The plant is also used as an antiseptic and insecticide. (ii) *Peristrophe bicalyculata* : Used as an antidote for snake bite.
9. Rubiaceae	*Rubia*	$\oplus$ ⚥ $K_{(4-5)}\ C_{(4-5)}\ A_4\ G_{\overline{(2)}}$	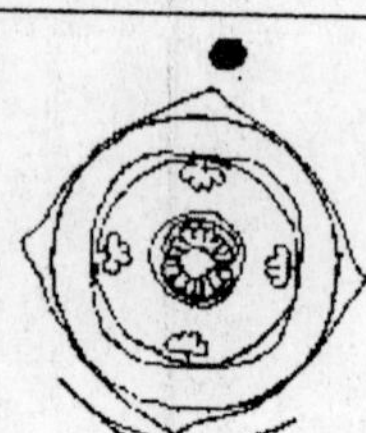 *Oldenlandia corymbosa*	From the point of view of economic important the family is of great value. (i) *Cinchona officinalis* : The tree yields *quinine* used for malarial fevers. (ii) *Rubia tinctorum* : The root of this plant yields a red dye which is used for dyeing cloth.
10. Labiatae or Lamiaceae		Br Brl % ⚥ $K_{(5)}\ \overline{C_{(5)}\ A_{4\ or\ 2}}\ \underline{G_{(2)}}$	 *Leucas lavendulacefolia*	The family is of little economic value. Some are ornamentals. Some plants yield essential oil, while some possess medicinal value. Some of these plants are — (i) *Ocimum sanctum (Tulsi)* : This is treated as holy plant by Hindus. The juice of the leaves is used as a stimulant and as a remedy for bronchitis, the seeds are used as a remedy for urinogenital troubles. The plant is used as an antidote for snake-bite. (ii) *Leucas zeylanica* : The fresh juice of leaves is used as a remedy for headache & cold. The flrs are also used for cough & cold. It is also applied to skin diseases.
11. Verbenaceae	*Verbena*	Br Brl % ⚥ $K_{(5)}\ C_{(5)}\ A_{2+2}\ \underline{G_{(2)}}$	 *Lantana camara*	This family is of great eco. importance and contains several imp. ornamental, medicinal and timber yielding plants. (i) *Tectona grandis* : The wood is used for making

Name of the family	Type genus	Floral formula	Floral diagram	Economic importance
				furniture, cabinet work and construction work. The bark yields a yellow dye which is used for staining basket. (ii) *Vitex negundo* : A shrub, grown for reclamation of forest land, the branches are used for manufacturing basket.
12. Scrophulariaceae	*Scoparia*	% ⚥ $K_{(4-5)} \overline{C_{(4-5)} A_{4 \text{ or } 5}} G_{\underline{(2)}}$	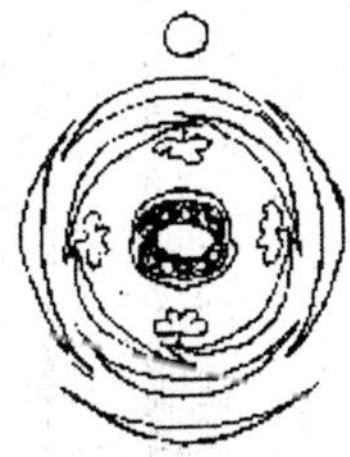 *Scoparia dulcis*	The family is of little oconomic value. Some plants are ornamental and some possess medicinal property. (i) *Linaria moroccana* : It is grown as an ornamental. It is used medicinally in diabetes. (ii) *Digitalls purpurea* : The leaves of thin plant are source of a drug which is used as a cardiac stimulant and tonic in heart diseases.
13. Asclepiadaceae	*Asclepias*	⊕ ⚥ $K_5 \overline{C_5 A_5} G_{\underline{(2)}}$	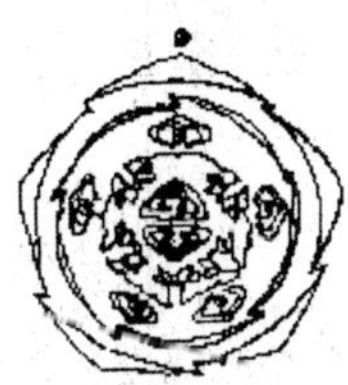 Calotropis procera syn. *Asclepias gigantea*	The family is of little economic value. Some plants are ornamental. Some have medicinal value. (i) *Asclepias kurassavica* : It is cultivated as ornamental but its root is used as a purgative and as a medicine for piles. (ii) *Calotropis procera* : The stem fibre is made into cordage. The stem and leaves are used medicinally in several ways.
14. Apocynaceae	*Nerium*	⊕ ⚥ $K_{5 \text{ or } (5)} C_{(5)} A_{10} G_{\underline{(3-5)}}$	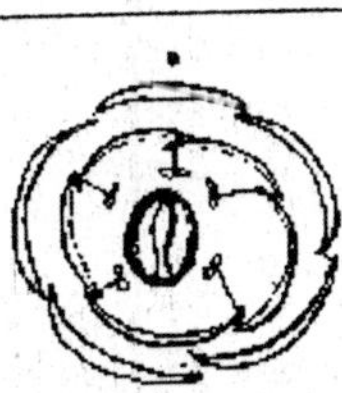 *Nerium indicum*	The family is of little economic value. Some of the plants are (i) *Rauwolfia serpentina* : The roots of this plant possess medicinal properties and are used in the treatment of hypertension, mental disorers etc.

Name of the family	Type genus	Floral formula	Floral diagram	Economic importance
				(ii) *Alstonia scholaris* : This small tree grown as ornamental but the bark of this plant can be used for diarrhoea & dysentry. Its latex is applied to ulcers.
15. Solanaceae	*Solanum*	⊕ ⚥ $K_{(5)}\ \overline{C_{(5)} A_5}\ G_{\underline{(2)}}$	 *Solanum torvum*	Economically the family is fairly important. It comprises of several plants of food value, medicinal, vegetables & ornamentals. Some of these are — (i) *Solanum tuberosum* : The tubers of this plant (potato) are used as vegetable & are also the source of starch. (ii) *Nicotiana tabacum* : The leaves are used for smoking & also contain alkaloids, which are used as insecticides. The oil obtained from the seeds is used for burning purposes & for manufacturing paints and varnishes. The oil cake is used as good manure.
16. Compositae or, Asteraceae	*Aster*	⊕ ⚥ $K_{Pappus}\ C_{(5)}\ A_{(5)}\ G_{\overline{(2)}}$	 Disc floret of *Helianthus annuus* 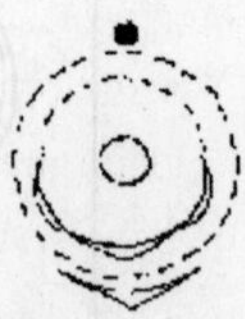 Ray floret of *H. annuus*	The family is of great economic value. Many plants possess medicinal & many have ornamental value such as — (i) *Helianthus annuus* : This is an ornamental large herb. The seeds yield an oil which is used in food, salad and burning purposes. The oil is also used in the manufacture of soap & varnish. (ii) *Calendula officinalis* : This is an ornamental herb grown for its beautiful flowers in gardens.

Family	Sub-family	Type genus	Floral formula	Floral diagram	Economic importance
17. Leguminosae or, Fabaceae	Caesalpinieae	*Caesalpinia*	% ⚥ $K_{5 \text{ or } (5)}$ $C_5\, A_{10}\, \underline{G_1}$	 *Caesalpinia pulcherrima*	Economically this sub-family is fairly important : (i) *Cassia alata* : The leaves are used for ringworms & skin diseases. Also grown as ornamental. (ii) *Cassia fistula* : This is an ornamental tree, the pulp of the fruits is used as a purgative and laxative.
	Papilionoideae		% ⚥ $K_{(5)}$ $C_{1+2+(2)}$ + $A_{1+(9)}\, \underline{G_1}$	 *Pisum sativum*	This family is of great economic value. (i) *Pisum sativum* : The seeds and buds are eaten as vegetable. They are quite rich in proteins. (ii) *Dalbergia sissoo* : The wood is used for making furniture and carving. The pulp is used for making writing & printing papers.

Symbol used to write floral formula

Br.	—	Bracteate.
Ebr.	—	Ebracteate.
⊕	—	Actinomorphic or regular flower.
% or + or Φ	—	Zygomorphic or irregular flower.
(⚥)	—	Bisexual flower.
♀	—	Female flower.
♂	—	Male flower.
K	—	Calyx.
C	—	Corolla.
P	—	Perianth.
A	—	Androecium.
G	—	Gynoecium.
$\underline{G}$	—	Superior ovary.
$\overline{G}$	—	Inferior ovary.
1, 2, 3, 4, ... ∝	—	Number of sepals, petals, stamens or carpels.
()	—	Fused or united.
$\overline{CA}$	—	Epipetalous condition, i.e., stamens attached to petals.

13 ECOLOGY

Q. 1. What is autoecology?

Ans. Ecological study of individual plant or animal species is called as autoecology.

Q. 2. What is synecology?

Ans. Ecological study of a group of plants i.e. community, is known as synecology.

Q. 3. What is an ecosystem?

Ans. It is a functional unit where biotic and abiotic components of the environment interplay.

Q. 4. What is a community?

Ans. A community is a group of population of different species in a given area.

Q. 5. Define food chain.

Ans. The transfer of food energy from the source in plant through a series of organisms eating and being eaten is called food chain.

Q. 6. Define food web.

Ans. In a particular ecosystem different kinds of food chain intermingle with one another and thus create a complicated process known as food web.

Q. 7. What is energy flow?

Ans. In an ecosystem the solar energy is the main source of all energy. The transformation and transmission of solar energy from one trophic level to another trophic level is known as energy flow.

Q. 8. What is greenhouse effect?

Ans. Due to the urbanisation and industrialisation, air pollutants like CO_2, NO_2, NH_4, CH_4 and CF-C_S etc increase. This glass like action of a act like a glass, which traps heat as the glass of a greenhouse.

Q. 9. What is global warming?

Ans. Certain gas pollutant like CO_2, NO_2, CH_4 and CFCs allow sunrays to pass through but then they absorb and radiate the heat back to earth thus heat the earth's surface and increase the global temperature. Due to this, gradual rise in sea level is caused by melting of snow which cause floods in coastal areas. This phenomenon is called global warming.

Q. 10. Define ozone hole.

Ans. Some chloroflurocarbons (CFCs) and halocarbonds (C×F×Brx) cause destruction of ozone shield in the startosphere region known as ozone hole which cause the entry of more UV rays causing harmful effects on living beings.

Q. 11. What do you mean by photochemical smog?

Ans. Two components in automobile exhaust (nitrogen oxide and hydrocarbon) combine in the presence of sunlight to produce new and more toxic substance (PAN and O_3) known as photochemical smog.

Q. 12. What is ecotype?

Ans. Ecotype is a population of individuals of a species which are genetically different.

Q. 13. What is ecotone?

Ans. There is generally a zone of transition, presenting a situation of special ecological interest between two different types of community, which is called ecotone.

Q. 14. What is plant succession?

Ans. Plant succession is the natural process by which the same locality becomes successively colonised by a different group of community of plants.

Q. 15. What do you understand by biomass?

Ans. Biomass indicates the total mass of organism which are present on the earth.

Q. 16. Define biosphere.

Ans. The portion of the earth in which ecosystem operates is called biosphere.

Q. 17. What is climax?

Ans. The final stage of a community is called climax which is stable and fixed.

Q. 18. What is ecological niche?

Ans. Taxonomic components in an ecosystem are termed ecological niche.

Q. 19. What is biogeochemical cycle?

Ans. More or less circular pathway through which the chemical elements circulate in the biosphere from environment to organism and back to the environment known as biogeochemical cycle. E.g. sulphur cycle, carbon cycle, nitrogen cycle.

Q. 20. What is physiologically dry soil?

Ans. In some soil as those of halophytic condition, although water is present in abundance, due to high degree of salinity, it is not available to plants. Such soil are referred as physiologically dry soil.

Q. 21. Define biome.

Ans. A major ecological community of organism maintained under a particular climatic zone called biome.

Q. 22. What is biota?

Ans. The flora and fauna of an area is known as biota.

Q. 23. What do you mean by decomposer in an ecosystem?

Ans. Microbes that obtain their nutrition from breakdown products of dead organic matter.

Q. 24. Define detritus food chain.

Ans. Dead organic matter, mainly of fallen leaves, are decomposed by detritus microbe – this is called detritus food chain.

Q. 25. What is ecesis?

Ans. The successful establishment of a species in a new area is called ecesis.

Q. 26. What is mutualism?

Ans. Mutualism is an association where both organisms are mutually benefited and are unable to survive in isolation.

Q. 27. What is in situ conservation?

Ans. In situ conservation means conserving the plants or animals within their natural habitat.

Q. 28. What is ex situ conservation?

Ans. Ex- situ conservation means conserving the plants or animals out of their natural habitat.

Q. 29. Mention the different components of an ecosystem.

Ans. Generally 4 essential components are noted in an ecosystem :

(i) Abiotic components.

(ii) Producer components.

(iii) Consumer components.

(iv) Decomposer.

Q. 30. What is eutrophication?

Ans. Extensive growth of microbes in water bodies consuming dissolved O_2, and addition of plant nutrients in excess causing decrease in the precentage of dissolved O_2 in water is called eutrophication.

Q. 31. What are pneumatophores?

Ans. In some halophytic plants problem of respiration is solved by negatively geotropic roots which stand vertically above saline water.

Q. 32. What is quadrate study?

Ans. The study of vegetation by square, rectangular or circular method of any particular spot is known as quadrate study.

Q. 33. What are non-degradable pollutants?

Ans. There are material and poisonous substances like aluminium cans, mercuric salts, DDT etc. that either can't degrade or degrade only very slowly in nature – these are known as non-degradable pollutants.

Q. 34. What are biodegradable pollutants?

Ans. Biodegradable pollutants are those that can be rapidly decomposed either by natural or "engine red process".

Q. 35. What do you mean by acid rain?

Ans. Due to some natural processes sometimes the acid of the air – like nitric acid, sulphuric acid – comes down on earth. As a result, acid is found in the rain water. This is called acid rain.

Q. 36. What is biodiversity?

Ans. Biodiversity is the total variety of life form on the earth's surface.

Q. 37. What is plankton flora?

Ans. Microorganisms which are free floating throughout their entire life cycle are known as plankton which have different category called plankton flora.

Q. 38. What do you mean by microclimate?

Ans. Microclimate refers to total combination of atmospheric factor which is due to local variation in the climate.

Q. 39. What do you mean by `Age and area hypothesis'?

Ans. According to this hypothesis, the occurence of diverse species and genera in a particular space is proportional to the age in the evolution.

Q. 40. What do you mean by endengered species?

Ans. The plant species which are at the verge of extinction are known as endengered species.

Q. 41. What is endemic species?

Ans. The species which are confined to a small area and beyond which their existence is not found are called endemic species.

Q. 42. What is pedology?

Ans. The study of soil development is known as pedology.

Q. 43. What is ecological pyramid?

Ans. It is the trophic structure of an ecosystem. Sometimes represented by trophic function. In ecological pyramid the producer level constitute the base while the successive level makes up the apex.

Q. 44. Define pyramid of biomass.

Ans. In any ecosysem if the food levels are arranged according to the biomass then a pyramid is formed and that pyramid is called pyramid of biomass. There is generally a gradual decrease in biomass of oganism at successive levels from the producer to the top. The pyramid is upright.

Q. 45. What is pyramid of energy?

Ans. In an ecosystem if the food levels are arranged according to the energy then a pyramid is formed and that pyramid is called pyramid of energy.

Q. 46. What do you mean by inverted pyramid?

Ans. When the value of biomass gradually shows an increase towards the apex of the pyramid, thus making the pyramid inverted.

It is found in case of parasitic ecosystem.

Q. 47. What are mangroves?

Ans. It is halophytic in nature. This vegetation grows and thrives best in soil containing high amount of salt. Mangrove vegetation is mainly composed of annual and perennial herb and woody shrubs and tree.

Q. 48. What is endemism?

Ans. The phenomenon of occurrence of species, genera or other group to a small area beyond which their existence is not found is known as endemism.

Q. 49. What are epiphytes?

Ans. Epiphytes are the plant that seek accommodation in other plants but do not absorb food from them. E.g. *Vanda* sp.

Q. 50. Name two greenhouse gasses.

Ans. Methane (CH_4), Chloroflurocarbon (CFCs).

Q. 51. What is cryopreservation?

Ans. Cryopreservation means storing a material at very low temparature. All biological functions cease at this temperature, hence growth and ageing are suspended.

Q. 52. Name a cryoprotective agent.

Ans. Dimethyl sulphoxide.

Q. 53. What is transect study?

Ans. It is a method of studying vegetation area. Transect is a continuous narrow strip of any length which gives a cross-section of the vegetation.

Q. 54. What are the anatomical feature of xerophytes?

Ans. (i) Sunken stomata present.

(ii) Multilayered epidermis.

(iii) Cuticle present.

(iv) Present of hairs.

Q. 55. What are the `rare' and `threatened' plant species?

Ans. The rare species with small population in the world. At present these species are not endengered. They are usually restricted within geographical area or habitats.

The term threatened is used in the context of conservation of the species which are in any one of the categories – endengered, rare etc.

Q. 56. What do you mean by biotic and abiotic pollutants?

Ans. The pollution happened by the living organisms is called biotic pollution and pollutants called biotic pollutants whereas when pollution occur by the non-living organisms these are called as abiotic pollutants.

Q. 57. Mention the different components of a pond ecosystem.

Ans. (i) Abiotic, (ii) Producer (phytoplankton), (iii) Consumer small fishes are primary consumer, large fishes are secondary consumer and the bacteria are decomposer.

Q. 58. What is edaphic factor?

Ans. Edaphic or soil factor constitute all conditions of soil regarded mineral and organic matter contents, capacity or retention of water, human content etc.

Q. 59. What is neoendemics?

Ans. According to this endemic taxa are supposed to have evolved only during the recent times and did not have sufficient time to extend their range of distribution.

Q. 60. What is BOD?

Ans. Biological Oxygen Demand (BOD) is the amount of oxygen required for biological oxidation by microbes in any unit volume of water.

Q. 61. Water is hydrosere?

Ans. A sere taking its origin in water is termed hydrosere.

Q. 62. Define pollution.

Ans. Pollution is an undesirable change in the physical, chemical or biological characteristics of our air, land and water that may or will harmfully affect human life, industrial process, living condition and cultural assets.

Q. 63. What is the aim of conservation of natural resources?

Ans. The real aim of conservation is –

(a) To ensure the preservation of a quality environment that considers aesthetic and recreational as well as product need, and

(b) to ensure a continuous yield of useful plants, animals and materials by establishing a balanced cycle of harvest and renewal.

Q. 64. What is nuddation?

Ans. In plants, succession development of a base area is known as nuddation.

Q. 65. What is frequency in community?

Ans.

$$\text{Frequency} = \frac{\text{Number of sampling unit in which the species occurred}}{\text{Total number of sampling units studied}} \times 100$$

Q. 66. What is adaptation?

Ans. Any feature of the organism or its part which is of definite significance of allowing that organism to exist under the condition of its habitat is called adaptation.

Q. 67. What is antibiosis?

Ans. The term `antibiosis' generally refers to the complete or partial inhibitor or death of one organism by another through the production of some substances.

Q. 68. What are indicator plants?

Ans. The organism, chiefly plants, species, communities or even systems serves as a measure or indicator of the environment. If plants serves as indicator then the plants are known as indicator plants. E.g. Water bloom is a indicator of water pollution. Here algae acts as pollution indicators.

Q. 69. What is physically dry soil?

Ans. Sandy soil due to bigger pore species are unable to hold much water and thus are well-drained. Such soils are known as physically dry soil.

Q. 70. What are the sources of water pollution?

Ans. (i) Sewage and other wastes

(ii) Industrial effluents

(iii) Agricultural discharges

(iv) Industrial wastes.

14 PHARMACOGNOSY AND ECONOMIC BOTANY

Q. 1. What is pharmacognosy?

Ans. Pharmacognosy is the objective study of crude drugs of plants and animals treated scientifically.

Q. 2. What is pharmacology?

Ans. Pharmacology is an outgrowth of materia medica. It is also an essential branch of medicine and related to the study of the action of drugs.

Q. 3. What do you mean by drug?

Ans. Drug means those substances – natural or synthetic – having therapeutic or medicinal properties and usually used as medicine or as ingredient of medicines.

Q. 4. What is crude?

Ans. Crude means any product which has not been advanced in value or improved in condition by grinding, crushing, distilling, evaporating, extracting by artificial mixing with other substance or by any other process.

Q. 5. What do you mean by organoleptic study?

Ans. Organoleptic study refers to evolution by means of organ of sense and includes the macroscopic appearance and the odour, test etc. of the drug.

Q. 6. What is adulteration?

Ans. Adulteration is the debasement of any article. It usually occurs when the drug is scarce or when its price is normally high, though there may not be any scarcity.

Q. 7. What are the main alkaloids of *Rauwolfia*?

Ans. The main alkaloids of *Rauwolfia* are :

(i) Reserpine, (ii) Rescinnamine, (iii) Ajmalinine, (iv) Ajmaline, (v) Serpentine, (iv) Serpentinine, (vii) Rauwolfinine.

Q. 8. Mention the major uses of *Rauwolfia serpentina*.

Ans. (i) Reserpine is mainly used for lowering of blood pressure and sedative influence.

(ii) The drug is also useful in gynaecological conditions.

(iii) Ajamline is used for the treatment of cardiac arrhythmias.

Q. 9. What are the main alkaloids of *Adhatoda* leaf?

Ans. The main alkaloids are :

(i) Vasicine, (ii) Vasicinone, (iii) 6-hydroxy vasicine, (iv) Volatile oil, (v) Adhatodic acid.

Q. 10. Mention the major uses of *Adhatoda* leaf.

Ans. (i) The drug has long been used as an expectorant, particularly in cough, chronic bronchitis, asthma.

(ii) Vasicinon is broncho-dilator whereas vasicine is broncho-constrictor.

(iii) The leaves are also useful as mild bronchial antipasmodic.

Q. 11. What do you mean by constituents of drug?

Ans. The chemical compounds which give plant and animal drugs their therapeutic properties are known as constituents of drug.

Q. 12. What are the main alkaloids of *Strychnos nux-vomica* seed?

Ans. Nux-vomica seed contains 1.5-5% of bitter indole alkaloid. Chief constituents of nux-vomica are – Strychnine, brucine, vomicine, α-colubrine, pseudostrychnine, strychnicine.

Q. 13. Mention the major uses of nux-vomica seed.

Ans. (i) The dried ripe seed are source of a drug nux-vomica which is used as tonic, stimulant and in the treatment of paralysis.

(ii) Powdered seed is used in chronic rheumatism.

(iii) Brucine is used as an alcohol denaturant.

Q. 14. What are the major alkaloids of *Cinchona succirubra?*

Ans. The major alkaloids are :

(i) Quinine, (ii) Quinedine, (iii) Cinchonine, (iv) Cinchonidine, (v) Hydrocinchonidine, (vi) Homocinchonidine.

Q. 15. Mention the major uses of *Cinchona succirubra* bark.

Ans. (i) Cinchona and it alkaloids has been used in the treatment of malaria fever.

(ii) Quinine is used to treat various cardiac arrhythms.

(iii) The bark is given chiefly as bitter stomachics and tonics.

Q. 16. What is retting of jute?

Ans. After harvesting, the bundles of the stem are taken in a stagnant pool or ditch and laid flat in water, arranged side by side. Another layer of bundle is placed on the top at right angles to the first. The surface of the bundle is covered with wood. This process is known as retting of jute.

Q. 17. What is rice-bran?

Ans. Rice bran, obtained during polishing brown rice for producing white rice.

Q. 18. What are the major uses of rice-bran?

Ans. (i) Rice-bran used as a liverstock and poultry feed.

(ii) Oil extracted from the bran, is used for cooking and in the manufacture of soap.

Q. 19. What is C.T.C. tea?

Ans. Besides, the orthodox method of tea processing a variation was introduced in the process of manufacturing tea, especially in North India. The variation consists of the use of a machine named `Crushing, Tearing and Curling machine (C.T.C.)

Q. 20. What do you mean by legg-cut tea?

Ans. The non-withered leaf is difficult to roll, the leaf is cut into small strips by using a chaff-cutter also known as legg-cut tea.

Q. 21. What are the alkaloids of tea leaf?

Ans. The principal alkaloids of tea leaf are caffeine, ortho-quinone, theoflavins, theophylline etc.

Q. 22. Mention the major uses of sugarcane.

Ans. (i) Sugarcane is an important source of sugar.

(ii) It is used in hair oil, medicine etc.

(iii) Cane-wax after extraction with solvent is used in manufacturing polishes, cosmetics and paper containers.

Q. 23. State the major uses of coconut.

Ans. (i) Coconut oil is used for cooking in South India.

(ii) The coconut oil used as hair oil and in the preparation of best quality soap, cosmetics, shampoo etc.

(iii) Oil cake used as fodder and manure.

Q. 24. What is the scientific name of *sal*?

Ans. The scientific name of *sal* is *Shorea robusta.*

Q. 25. What is the family of *sal*?

Ans. The family of *sal* is Dipterocarpaceae.

Q. 26. What are the major uses of *sal*?

Ans. (i) *Sal* timber is used for various purposes, e.g. for piles, beam, planking and railing or bridges.

(ii) It is also used for beams, doors, window frames of houses, for gun-carriages.

(iii) A type of whitish, aromatic, tansparent resin known as `*lal dhuna*' is obtained from the tree.

Q. 27. What is lint and fuzz?

Ans. The lint is a type of hair which is grown long; in cross-section there is a lumen due to reduced deposition of cellulose.

The fuzz are short hairs, the lumen is almost closed due to cellulose deposition. The hairs are strongly attached to the seed.

Q. 28. Name a high yielding variety of rice.

Ans. *Oryza sativa.*

Q. 29. What is the botanical name of white jute?

Ans. The botanical name of white jute is *corchorus capsularis.*

Q. 30. What is the full form of CRRI?

Ans. The full form of CRRI is Central Rice Research Institute.

SOME ECONOMICALLY IMPORTANT PLANTS

1. Cereals

No.	Common name	Botanical name	Family	Plant part used
1.	Wheat	*Triticum aestivum*	Gramineae	Seed (Fruit)
2.	Rice	*Oryza sativa*	Gramineae	Seed (Fruit)
3.	Maize	*Zea mays*	Gramineae	Seed (Fruit)
4.	Bajra	*Pennisetum glaucum*	Gramineae	Seed (Fruit)
5.	Jawar	*Sorghum vulgare*	Gramineae	Seed (Fruit)

2. Pulses and Legumes

No.	Common name	Botanical name	Family	Plant part used
1.	Arhar	*Cajanus cajan*	Leguminosaè	Seed
2.	Gram	*Cicer arietinum*	Leguminosae	Seed
3.	Groundnut	*Arachis hypogea*	Leguminosae	Seed
4.	Moong	*Phaseolus aureus*	Leguminosae	Seed
5.	Pea	*Pisum sativum*	Leguminosae	Seed
6.	Soyabean	*Glycine max*	Leguminosae	Seed
7.	Urd	*Phaseolus mungo*	Leguminosae	Seed

3. Fibre-Yielding Plants

No.	Common name	Botanical name	Family	Plant part used
1.	Cotton	*Gossypium* sp.	Malvaceae	Seed coat
2.	Flax	*Linum usitatissimum*	Linaceae	Stem
3.	Jute	*Corchorus capsularis* *Corchorus olitorius*	Teliaceae	Stem
4.	Patsan	*Hibiscus cannabinus*	Malvaceae	Stem
5.	Sun hemp	*Crotolaria Juncea*	Papilionaceae	Stem

4. Medicinal Plants

No.	Common name	Botanical name	Family	Plant part used
1.	Aconite	*Aconitum heterophyllum*	Rannunculaceae	Root
2.	Bellodona	*Atropa bellodona*	Solanaceae	Root
3.	Serpentine or Sarpagandha	*Rauwolfia serpentina*	Apocynaceae	Root
4.	Cinnamon	*Cinnamomum zeylanicum*	Lauraceae	Bark
5.	Quinine	*Cinchona officinalis*	Rubiaceae	Bark
6.	Ephedrine	*Ephedra gerardiana*	Gnetaceae	Stem & Woods
7.	Katha	*Acacia catechu*	Mimosaceae	Stem
8.	Saffron	*Crocus sativus*	Iridaceae	Flower
9.	Opium	*Papaver somniferum*	Papaveraceae	Fruits
10.	Azadiractin	*Azardirachta indica*	Meliaceae	Fruits & Seeds
11.	Worm Seed (Kirmala)	*Artemisia martinima*	Rubiaceae	Flower
12.	Withania	*Withania somnifera*	Solanaceae	Root
13.	Ipecal	*Cephaelis ipecacuanha*	Rubiaceae	Rhizome
14.	Isabgol	*Plantago ovata*	Plantijinaceae	Seeds
15.	Ginseng	*Panax schinseng*	Araliaceae	Root
16.	Fox Glove	*Digitalis purpurea*	Scophulariaceae	Leaves

SOME PLANTS YIELDING FATTY OILS

No.	Common name	Botanical name	Family	Plant part used
1.	Coconut/Nariyal	*Cocos nucifera*	Palmae	Endosperm
2.	Mustard/Sarson	*Brassica compestris (Vas. sarson)*	Cruciferae	Seed
3.	Ground nut/Moongphali	*Arachis hypogaea*	Papilionaceae	Seed
4.	Safflower/Kusum	*Carthamus tinctorius*	Compositae	Seed
5.	Sesame, Gingelly/Til	*Sesamum indicum*	Pedaliaceae	Seed
6.	Soyabean	*Glycine max*	Papilionaceae	Seed
7.	Sunflower/Surajmukhi	*Helianthus annuus*	Compositae	Seed
8.	Oil palm/African oil palm	*Elaeis guineensis*	Palmae	Fruit pulp and kernel
9.	Castor/Arandi	*Ricinus communis*	Euphorbiaceae	Seed
10.	Cotton/Kapaas	*Gossypium spp.*	Malvaceae	Seed
11.	Linseed/Alsi	*Linum usitatissimum*	Linaceae	Seed
12.	Olive/Jaitoon	*Olea europea*	Oleaceae	Fruit

IMPORTANT RESIN-YIELDING PLANTS

No.	Name	Botanical Name	Family	Characteristics
1.	Asafoetida (Hing)	*Ferula asafoetida*	Umbelliferae	Antihelmintic, used in perfumery and flavouring agent
2.	Kaladamar	*Shorea tumbuggaia*	Dipterocarpaceae	Resin from stem, used in incense
3.	Lacquer	*Rhus vernicifera*	Anacardiaceae	Applied as varnish
4.	Turpentines	*Plnus roxburghi*	Pinaceae	Varnishes, paints
5.	White damar	*Veteria indica*		Used in incense, paints, varnishes

BEVERAGES

1. Non-Alcoholic

No.	Common Name	Source plant/ Botanical name	Family	Plant part used
1.	Tea	*Thea sinensis*	Theaceae	Young leaves and short tips
2.	Coffee	*Coffea arabica*	Rubiaceae	Seed
3.	Cocoa	*Theobroma cacoas*	Sterculiaceae	Seed

2. Alcoholic

No.	Common Name	Source plant/ Botanical name	Family	Characteristics
1.	Beer	*Hordeum vulgare*	Gramineae	Barley malt used, 3-6% alcohol
2.	Brandy (from grapes)	*Vitis vinifera*	Vitaceae	Fermented and distilled juice, 60-70% alcohol
3.	Rum (from molasses)	*Saccharum officinarum*	Gramineae	40% Alcohol
4.	Whisky	*Cereals & Potatoes*		Distilled alcohol, 50% alcohol

IMPORTANT COMMERCIAL WOODS

No.	Name	Common name	Botanical name	Family	Salient features
1.	Indian red wood	Sheesham	*Dalbergia sisoo*	Papilionaceae	Finest wood for furniture. Also used for railway sleepers etc.
2.	Pine wood	Chir	*Pinus roxburghi, Pinus spp.*	Pinaceae	Used in match industry, railway sleepers and packing boxes
3.	Teak	Saugaun	*Tectona grandis*	Virbenaceae	House construction, ships, bridges, boats etc.
4.	Cedar	Deodar	*Cedrus deodara*	Pinaceae	Very strong Indian soft wood
5.	Mahogany	—	*Swietenia mahogonie*	Meliaceae	Light reflecting quality wood. Used in aeroplane propellers, ship building and high class furniture
6.	Sal	Sal	*Shorea robusta*	Dipterocarpaceae	Next to teak, frequently used in construction work
7.	Toon	Toon	*Cedrella toona*	Meliaceae	Used for packing cases, furniture etc.
8.	Ebony	—	*Diospyrons abenum*	Ebenaceae	Used in decoration work, Piano keys, walking sticks, flutes etc.
9.	Birch	Bhojapatra	*Betula alnoides*	Betulaceae	Furniture and plywood
10.	Balsa	—	*Ochroma longopus*	—	Lightest commercial wood
11.	Sandalwood	Chandan	*Santalum album*	Santalaceae	Used in religious ceremonies

SOME IMPORTANT FUMITORIES AND MASTICATORIES

No.	Name	Source Plant/ Botanical name	Family	Characteristics
1.	Tobacco	*Nicotiana tabacum*	Solanaceae	Leaves contain **nicotine,** mild stimulant, causes lung cancer and atherosclerosis of coronary arteries; accelerates heart beat, increases hypertension and bronchial cough
2	Kola	*Cola nitida*	Sterculiaceae	Seeds used as masticatory, contain glocoside kolanin and alkaloid caffeine
3.	Areca nut (Supari)	*Areca catechu* *Betel nut palm*	Palmae	**Endosperm** of the nut used as masticatory along with betel (*paan*); used as vermifuge for tapeworm in veterinary practice
4.	Betal, Paan	*Piper betel*	Piperaceae	The leaves provide *paan*
5.	Coca	*Erythroxylon coca*	Erythroxylaceae	Leaves contain **cocaine**; it is chewed with morphine or heroin — **speed ball**; Acts on central nervous system causing **psychic exaltation**; reduces appetite, physical and mental deterioration leads to death
6.	Hemp Indian	*Cannabis sativa*	Cannabinaceae	**Female flowers** used for extraction of hallucinogenic narcotics *hashish, charas, marijuana, bhang, ganja etc.* causes addiction.
7.	Opium	*Papaver somniferum*	Papaveracoae	Latex from unripe capsules yield alkaloid

Spices and Condiments

No.	Common Name	Botanical Name	Family
1.	Dhania/Coriander	*Coriandrum sativum*	Umbelliferae
2.	Adrak/Ginger	*Zingiber officinale*	Zingiberaceae
3.	Heeng	*Ferula asafoetida*	Umbelliferae
4.	Red Pepper	*Capsicum annum*	Solanaceae
5.	Zira/Caraway	*Cuminum cyminum*	Umbelliferae
6.	Saunf/Fennel	*Foeniculum vulgare*	Umbelliferae
7.	Elaichi/Cardamom	*Eletteria cardamonum*	Zingiberaceae
8.	Peppermint	*Mentha piperita*	Labiatae
9.	Laung/Clove	*Syzygium aromaticum*	Myrtaceae
10.	Vanilla	*Vanilla planifolia*	Orchidaceae
11.	Kali Mirch/Black Pepper	*Piper nigrum*	Piperaceae
12.	Cassia/Tejpat	*Cinnamomum tamala*	Lauraceae
13.	Jaiphal/Nutmeg	*Myristica fragrans*	Myristiceae
14.	Haldi/Turmeric	*Curcuma domestica*	Zingiberaceae

Essential Oils

No.	Name	Botanical Name	Family	Characteristics
1.	Jasmine oil/Juhi	*Jasminum auriculatum*	Oleaceae	Fragrant flowers yield Jasmine oil used in perfumes, perfumed oils and attars
2.	Khas Khas Oil/ Khas Khas	*Vetiveria zizanioides*	Gramineae	Roots yield khus khus oil used in perfumes, soaps, *sherbets*; roots used for mats for air-coolers
3.	Lavender	*Lavandula officinalis*	Labiatae	Flowers yield oil for perfumes, toilet soaps, toilet powders, Lavender water etc.
4.	Rose oil/Gulab	*Rosa damascena*	Rosaceae	Petals yield oil of roses, used in rose water attar; petals used for gulkand
5.	Sandalwood oil/ Chandan	*Santalum album*	Santalaceae	Heartwood yields oil, oil used for toilet soaps, face creams, perfumery and also religious ceremonies
6.	Camphor/Kapoor	*Cinnamomum camphora*	Lauraceae	Wood yields Camphor (Camphor gum) used in perfumery and medicines
7.	Lemongrass oil	*Cymbopogon citratus*	Gramineae	Leaves yield oil which contains citral; used in Ionone, soaps, cosmetics etc.
8.	Eucalyptus oil	*Eucalyptus*	Myrtaceae	Dried leaves yield oil used for perfumery, treatment of asthma and bronchitis
9.	Cedarwood oil	*Juniperus macropoda*	Pinaceae	Heartwood yields oil, oil used for perfumery, clearing agent in the preparation of microscopic slides
10.	Peppermint oil	*Mentha piperita*	Labiatae	Leaves yield oil used in soap and perfume industry
11.	Champaka oil	*Michelia champaca*	Magnoliaceae	Flowers yield oil used in perfumery
12.	Clove oil	*Syzygium aromaticum*	Myrtaceae	Unopen flower buds yield oil used in perfumes and medicines

15 PLANT PHYSIOLOGY

Q. 1. What is aerobic respiration?

Ans. Aerobic respiration is the biological process by which reduced organic compounds are metabolised and subsequently oxidised in a controlled manner in the presence of oxygen. During this respiration, free energy is released and incorporated into a form of ATP :

$$C_6H_{12}O_6 + 6O_2 + 6H_2O \longrightarrow 6CO_2 + 12H_2O + 686 \text{ Kcal}$$

Q. 2. What is glycolysis?

Ans. Glycolysis is a series of reactions of respiration carried out by a group of soluble enzymes located in the cytosol. Chemically, glucose is partly oxidised to produce two molecules of pyruvate (a three carbon compound) and little ATP.

Q. 3. What is TCA cycle?

Ans. In aerobically respired organisms pyruvate is oxidised completely to CO_2 and a considerable amount of reducing power, $NADH^+$, $FADH^+$, are generated i.e. called TCA cycle (Tri carboxylic acid cycle). It occurs in mitochondria matrix.

Q. 4. What do you mean by electron transport occurs?

Ans. In the higher organisms the electron transport system (ETS) consists of a collection of electron transport through the inner membranes of mitochondia. This is responsible for release of a large amount of free energy. This system also found in chloroplast of plants.

Q. 5. What do you mean by substrate level phosphorylation?

Ans. In some cellular reaction the synthesis of ATP from ADP invovles the direct transfer a phosphate group from a substrate molecule to ADP, to form ATP is called substrate level phosphorylation.

As an example in glycolysis, 1, 3 Bi PGA $\xrightarrow[\text{ADP + Pi} \;\curvearrowright\; \text{ATP}]{\text{Phosphoglycorate kinase}}$ 3PGA

Q. 6. What is gluconeogenesis?

Ans. The synthesis of glucose through reversal of the glycolytic pathway is known as gluconeogenesis. Gluconeogenesis is not common in plants, but occurs frequently in animals.

Q. 7. Who first proposed the TCA cycle?

Ans. The German-born British biologist Hans A. Krebs, in 1937, reported the TCA cycle. Hans Krebs was awarded the Nobel Prize in medicine and physiology in 1953.

Q. 8. What do you mean by floating respiration?

Ans. Blackman termed the respiration of carbohydrate, fat and organic acid as floating respiration.

Q. 9. What is protoplasmic respiration?

Ans. In a respiration when protein is used as respiratory substrate it is called protoplasmic respiration.

Q. 10. What is oxidative anabolism?

Ans. In respiration all the intermediates of glycolysis are not converted into pyruvic acid. Some of them build back the carbohydrates and the phenomenon is called oxidative anabolism.

Q. 11. What organisms are considered as facultative anaerobes?

Ans. The microorganisms which can exist in presence as well as absence of oxygen are described as facultative anaerobes.

E.g. Yeast, member of Ascomycetes fungi.

Q. 12. What do you mean by oxidative decarboxylation?

Ans. In respiration, the conversion of pyruvic acid to acetyl CoA probably occurs in perimitochondrial space and is called oxidative decarboxylation. It is done by its cofactors.

Q. 13. Which enzymes of Kreb cycle are found in cytoplasm?

Ans. α-ketoglutaric dehydrogenase and succinic dehydrogenase.

Q. 14. What is the full form of FMN?

Ans. Flavin mononucleotide.

Q. 15. In aerobic respiration how much energy is obtained from one molecule of ATP?

Ans. @ 1 ATP = 7.6 Kcal.

Q. 16. Who first discovered the Hexos Monophosphate Sunt pathway?

Ans. Harecker et al in 1951.

Q. 17. What is Pasteur Effect?

Ans. Under high oxygen concentration, oxidative phosphorylation occurs and glycolytic phosporylation decreases. This inhibition of glycolysis by oxygen concentration was observed by Louis Pasteur (1861) in his studies alcoholic fermentation and, hence, this phenomenon is called Pasteur Effect.

Q. 18. What do you mean by extinction point?

Ans. In aerobic respiration the concentration of O_2 is described as extinction point.

Q. 19. What is salt-induced respiration?

Ans. If a plant tissue is transferred from water to inorganic salt solution then the respiration rate increases. This phenomenon is called salt induced respiration.

Q. 20. Name some respiratory inhibitors.

Ans. Azides, cyanides, indoacetates, carbon monoxide inhibit respiration even in very low concentration.

Q. 21. What will happen when the rate of respiration becomes more than the rate of photosynthesis?

Ans. Plants gradually die of starvation.

Q. 22. Which apparatus is used to measure respiration?

Ans. Respirometer.

Q. 23. How much ATPs are produced during EMP pathway?

Ans. 8ATPs are produced when develops 1 molecule glucose converts into 2 molecules of pyruvic acids.

4 ATPs are produce when 1 molecule pyruvate develops.

Q. 24. Cut surface of fruits and vegetables often become dark – why?

Ans. At the site of cut surface oxidation of tannic acid occurs in presence of trace amount of iron from the knife.

Q. 25. How many ATP molecules will be produced during the production of one molecule of acetyl CoA from one mole of pyruvic acid?

Ans. 2 ATPs.

Q. 26. What do you mean by Enter-Daudoroff pathway?

Ans. Certain bacteria such as *Pseudomonas saccharophila, P. fluorescens,* lack of phosphofructokinase enzyme. They cannot degrade glucose by glycolytic process. They follow yet another pathway of sugar breakdown called Enter-Daudoroff pathway.

Q. 27. What is cytosol?

Ans. Enzymes of glycolysis found in the soluble portion of cytoplasm. This soluble protion of cytoplasm is called cytosol.

Q. 28. How many ATP molecules are produced during Krebs cycle?

Ans. 13 molecules ATP.

Q. 29. Write down the pathway of respiration in which net gain of ATP is 2 or 0.

Ans. α if from glucose there form 2 pyruvic acid, from that two ethyl alcohol or two lactic acid formed – then net gain of ATP is 2.

β if from glucose only one pyruvic acid forms and then one molecue of ethyl alcohol or lactic acid develops – then net gain of ATP is 0.

Q. 30. What is R.Q.?

Ans. In respiration the ratio between the volume of CO_2 given out and oxygen taken in simultaneously by a given weight of the tissue in a given period of time at standard temperature and pressure is called respiratory quotient.

RQ is volume of CO_2 evalued/volume of O_2 evalued.

Q. 31. What do you mean by bioluminescence?

Ans. Bioluminescence means production of light by the living organisms i.e. energy appears in form of light; it is also called "fox fire".

Among the plants this phenomenon is found in many cases of fungi, *Armillaria mellea, Collybia tuberosa, Panus incandescens, Polyporushanedi;* Bacteria *Beggiatoa* and some algae *Noctiluca* :

$$\text{ATP + Luciferin} \xrightarrow{O_2 + \text{ions}} \text{Luciferin–ATP complex.}$$

$$\text{Luciferin–ATP complex} \xrightarrow{\text{Luciferase}} \text{oxidised luciferin + ADP} + H_2O + \text{Pi + Light}$$

Q. 32. Where RQ value is 0 and infinity?

Ans. RQ value is 0 in xerophytic i.e. succulent plants. It is infinity in anerobic respired organisms.

Q. 33. What do you mean by redox potential?

Ans. The tendency for electron transfer and acceptance is called as redox potential (E^0). The term redox has been derived from reduction oxidation. The redox potential is the ratio of reduced form to the oxidised form and is expressed in volts or millivolts.

Q. 34. What do you mean by P/O ratio in respiration?

Ans. The ratio is the number of mols of ATP generated per atom of oxygen utilised in respiration; hence it is used as an index of oxidative phosphorylation. In 1941, S. Ochoa obtained the first reliable measurement of the P/O ratio.

Q. 35. What do you mean by external and internal respiration?

Ans. The physical process by which the living organisms take up oxygen and eliminate carbon dioxide through their body surface or specialised respiratory organs is called external respiration. The biochemical process of oxidation of food that occurs within living cells for releasing energy is called internal respiration or cellular (or tissue) respiration.

Q. 36. Which parts of the cell are involved in aerobic and anaerobic respiration?

Ans. Aerobic respiration ⟶ Cytosol and mitochondria

Anaerobic respiration ⟶ cytosol.

Q. 37. Where does citric acid cycle take place?

Ans. Citric acid cycle (Krebs cycle) occurs in the mitochondrial matrix.

Q. 38. Name the components of electron transport system.

Ans. Pyridinoproteins (NAD/$NADH_2$ & NADP/$NADPH_2$), Flavoproteins (FAD/$FADH_2$ & FMN/$FMNH_2$), Coenzyme Q and Cytochromes.

Q. 39. How many mols of ATP are generated from reoxidation of each mol of $NADH_2$ & $FADH_2$ through the electron transport system?

Ans. From $NADH_2$ – 3 mols of ATP

From $FADH_2$ – 2 mols of ATP

Q. 40. How is O_2 utilised in cellular respiration?

Ans. In cellular respiration, O_2 is used as the final oxidant in the electron transport systems i.e. the final phase of aerobic respiration, for oxidising the H atoms to H_2O by the action of the enzyme cytochrome oxidose.

Q. 41. What do you mean by beta oxidation of fatty acids?

Ans. β oxidation is the preliminary process of fatty acid oxidation in which fatty acids are oxidised at β carbon so that the last two carbon atoms from the carboxyl end of the fatty acid are split off as acetyl CoA rendering the fatty acid shorter by a two carbon unit.

Q. 42. Where does β-oxidation occur within cells?

Ans. β oxidation occurs in the mitochondrial matrix of cell.

Q. 43. What are homolactic and heterolactic fermentation?

Ans. Homolactic fermentation means the process of fermentation of sugars in which lactic acid is the sole end product.

Heterolactic (or mixed lactic) fermentation is that type of fermentation of sugars in which lactic acid is produced along with other end-products like ethanol, CO_2 and other organic compounds.

Q. 44. How many ATP molecules are obtained from fermentation of 1 molecule of glucose?

Ans. 2 molecules of ATP.

Q. 45. Why muscle cells and mature RBCs of man respire anaerobically?

Ans. The human skeletal muscle and RBCs respire anaerobically because they lack mitochondria in which aerobic respiration occurs. In these tissues, glucose is broken down by anaerobic glycolysis into two molecules of lactic acid.

Q. 46. What is photorespiration?

Ans. Uptake of atmospheric O_2 with a concomitant release of CO_2 by illuminated liquid. Molecular oxygen series as substrate for rubisco and the formed 2-phosphoglycolate enters the photorespiratory carbon oxidation cycle. The activity of the cycle recovers some of the carbon found in 2-phosphoglycolate, but some is list to the atmosphere. E.g. – Paddy, tobacco, soyabean etc.

Q. 47. Who first proposed the term photosynthesis?

Ans. Barnes, in 1898, first proposed the term photosynthesis and considered it as a light-dependent carbon assimilation process in green plants.

Q. 48. What is the photosynthetic unit?

Ans. The total amount of pigment molecules within the quantasome (approximately 400) absorbs a fixed amount of solar energy to carry out the photochemical light phase and is referred to as phtosynthetic unit.

Q. 49. Why photosynthesis is a redox reaction?

Ans. In photosynthesis CO_2 is reduced to glucose and H_2O is oxidised to O_2 – so it is a redox reaction.

Q. 50. Give the chemical reaction of photosynthesis.

Ans. $6CO_2 + 12H_2O \xrightarrow[\text{enzymes}]{\text{Light, chlorophyll}} C_6H_{12}O_6 + 6O_2 + 6H_2O$

Q. 51. What is action spectra of photosynthesis?

Ans. The region of the visible spectra in which photosynthesis occurs efficiently is known as action spectra. It occurs in 350 nm to visible to 680 or 750 nm light, i.e. visible light.

Q. 52. What is red drop?

Ans. The exposure of chlorophyll molecule to red and far red region (i.e. above 680 nm) of the visible spectrum reduces photosynthesis i.e. reduces production of $(CH_2O)_n$ in photosynthesis, called red drop.

Q. 53. What is Emerson's enhancement effect?

Ans. The exposure of chlorophyll molecule to alternation of long wave length of light, i.e. red light with short wave length of light, i.e. green light increases the rate of photosynthesis; this is Emerson's enhancement effect.

Q. 54. What is inductive resonance?

Ans. The transfer of solar energy from one pigment molecule to the other is known as inductive resonance.

Q. 55. What is triplet state?

Ans. A chlorophyll molecule reaches the metastable triplet state under the effect of blue light – it has a half life period of 10^{-3} seconds.

Q. 56. What is Z-scheme?

Ans. In light phase the noncyclic photo-phosphorylation scheme joining the PSI and PSII (pigment system) is called the Z-scheme. This is because of its `Z' – like configuration.

Q. 57. What are the end products of light reaction?

Ans. ATP, $NADPH_2$, O_2, H_2O.

Q. 58. What are the oxidising and reducing agents in photosynthesis?

Ans. Oxidising agent – CO_2

Reducing agent – H_2O

Q. 59. What are the necessary components of electron transport process in the light reaction of photosynthesis?

Ans. Quinone, plastoquinone, cyt b, cyt f, (cytochrome), ferredoxin, NADP, ADP.

Q. 60. What is solarisation?

Ans. The process of inhibition of photosynthesis due to intense light, which, in presence of O_2, causes bleaching of chloroplast.

Q. 61. What do you mean by Hill reaction?

Ans. In light phase the breaking of water molecules by isolated exiting chloroplasts in illuminated condition to produce H^+ & OH^- :

$$H_2O \longrightarrow H^+ + OH^-$$

This reaction was first time proposed by Robin Hill – so it is called Hill reaction.

Q. 62. What is Blackman's reaction?

Ans. Dark phase's reaction of photosynthesis by which CO_2 is fixed to form hexose molecule is known as Blackman's reaction :

$$6CO_2 + 12NADPH_2 \longrightarrow C_6H_{12}O_6 + 12NADP + 6H_2O$$

Q. 63. Why O_2 is not given off in bacterial photosynthesis?

Ans. Photosynthetic bacteria like green sulphur bacteria utilises other inorganic compounds like H_2S instead of H_2O as the electron donor, hence they liberate elemental sulphur instead of O_2.

Q. 64. What is C_4 cycle?

Ans. The process of carbon fixation in monocot plants by which CO_2 is absorbed by phosphoenol pyruvic acid (PEPA) to produce oxalo acetic acid and continue a cycle. OAA is the first stable compound which has 4 carbon so thin cycle is called C_4 cycle.

Q. 65. What is C_2 cycle?

Ans. The fixation of CO_2 by RUDP oxygenase in presence of light, high concentration of O_2 in the photosynthetic cells forming phosphoglycolic acid. It is also called photorespiration.

Q. 66. What is CAM?

Ans. In some xerophytic plants the process of crassulacean acid metabolism is followed by which fixation of CO_2 forms malic acid during night, which undergoes decarboxylation during daytime producing CO_2, keeping the stomata closed.

Q. 67. What is kranz anatomy?

Ans. In monocot C_4 plants, the vascular bundles of leaves remain surrounded by large chloroplast containing cells called bundle sheath cells. This type of anatomical structure is called Kranz anatomy.

Q. 68. Name two photosynthetic bacteria.

Ans. *Rhodopseudomonas* sp, *Chromatium* sp.

Q. 69. Name two chemosynthetic bacteria.

Ans. *Thiobacillus* sp., *Beggiatoa* sp.

Q. 70. What do you mean by antenna pigment?

Ans. A phtosynthetic unit consists of about 400 nm light harvesting chlorophyll molecules and a trapping centre. These tightly packed light harvesting chlorophyll molecules which allow energy migration by resonance transfer is called antenna pigment.

Q. 71. What are two special characteristics of CAM plants?

Ans. (i) In CAM plants stomata remains open at night and closed during day.

(ii) CAM plants are fleshy and succulent, so they can preserve water.

Q. 72. What is quantum requirement of photosynthesis?

Ans. In photosynthesis the number of protons required to generate one molecule of oxygen is the quantum requirement of photosynthesis. It ranges from 8 to 10.

Q. 73. Why Calvin cycle is called C_3 cycle?

Ans. From radioactive studies, the stable compound formed in dark phase of photosynthesis is found to be a 3-carbon compound known as phosphoglyceric acid (PGA), which contains 3-carbon atoms – hence the name "C_3 cycle"

Q. 74. Which cycle is called the Calvin-Bassham cycle?

Ans. The C_3 cycle of dark phase in photosynthesis is the Calvin-Bassham cycle. It is also termed as reductive pentose phosphate cycle, or Calvin-Benson cycle.

Q. 75. What is termperature coefficient?

Ans. The effect of temperature on the,rate of biological process is coveniently termed temperature coefficient or, more commonly Q_{10}. It is the ratio of the reaction rate (velocity) at 10°C higher by the velocity at particular temperature :

$$\text{Temperature coefficient } (Q_{10}) = \frac{\text{Velocity at } (T^0 + 10^0)}{\text{Velocity at } T^0}$$

Q. 76. Write down the structure of chlorophyll.

Ans.

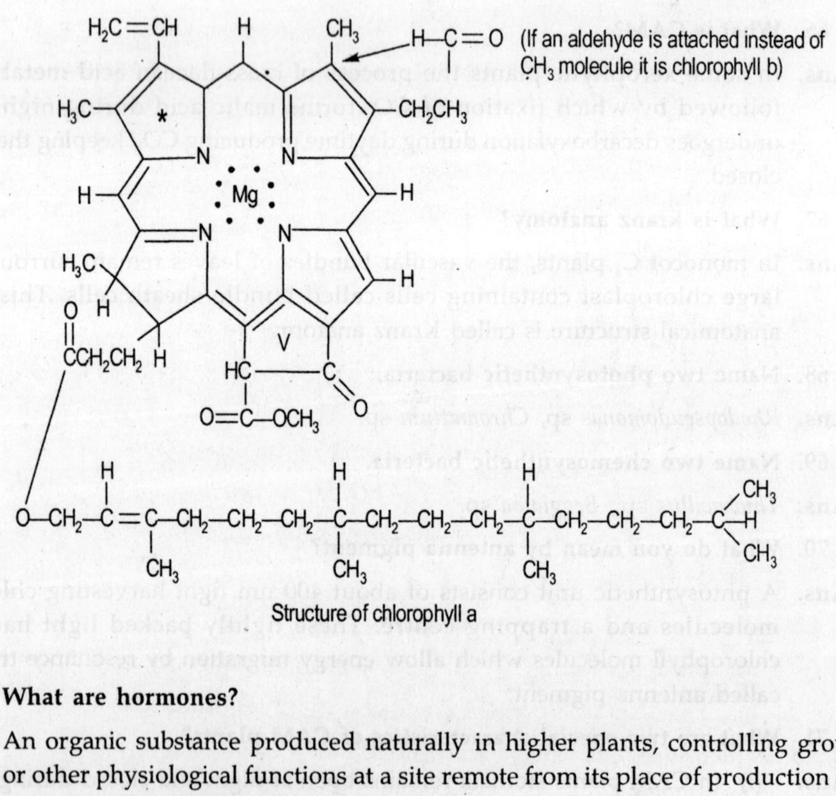

Structure of chlorophyll a

Q. 77. What are hormones?

Ans. An organic substance produced naturally in higher plants, controlling growth or other physiological functions at a site remote from its place of production and active in minute amounts.

Q. 78. What do you mean by auxins?

Ans. Auxin is derived from the Greek word "*auxein*" meaning "to grow" or "increase".

Q. 79. What are auxins?

Ans. Auxins can be defined as compounds which, in very low concentration promote the enlargement of newly formed cells. Chemically it is indol-3yl acetic acid, IAA.

Q. 80. Mention some naturally occuring auxins and their structure.

Ans.

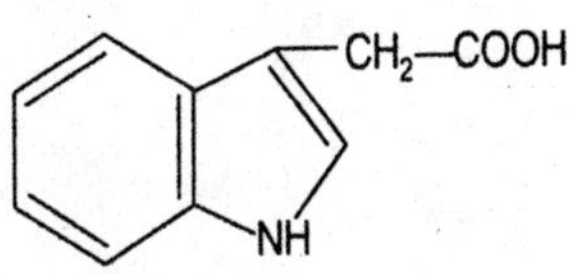

Indole acetic acid

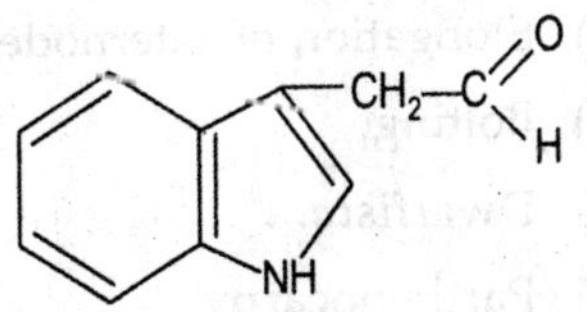

Indoleacetaldehyde

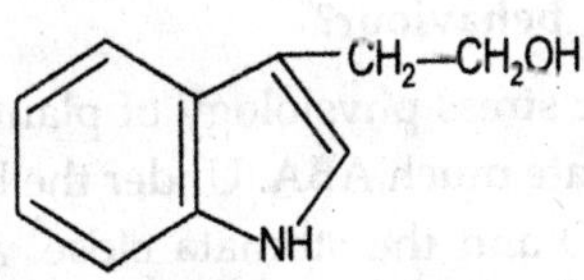

Indole ethanol

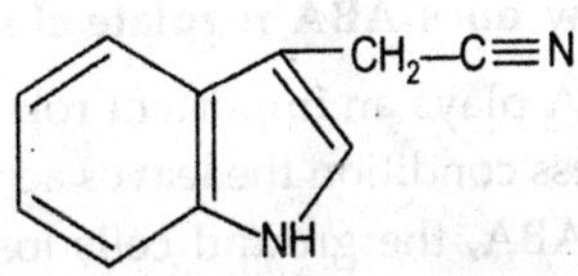

Indoleacetonitrile

Q. 81. Mention some artificially synthesised auxin hormones.

Ans. Naphthalene acetic acid (NAA), Indole butyric acid, and 2, 4-Dichlorophenoxy-acetic acid (2, 4-D) :

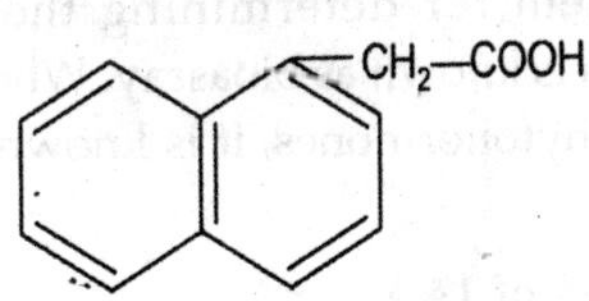

α naphthalene acetic acid (NAA)

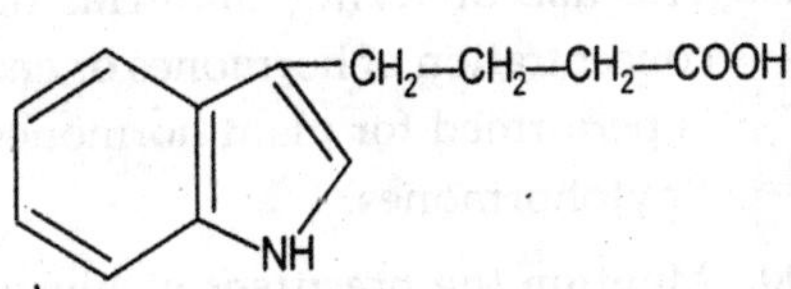

Indole butyric acid

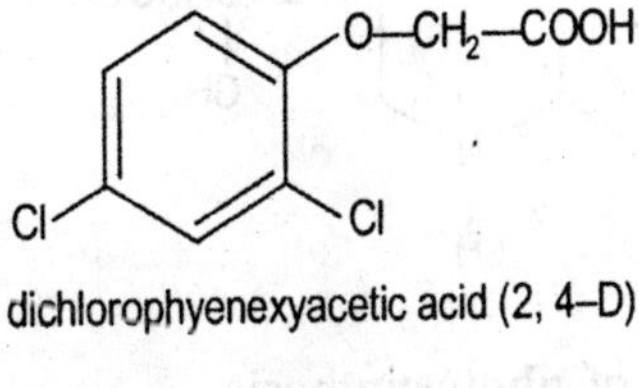

2, 4 dichlorophyenexyacetic acid (2, 4–D)

Q. 82. What is the role of auxins in plants.

Ans. (i) Auxins are employed to induce rooting, parthenocarpy, flowering.

(ii) They can also prevent abscission of leaves, fruits and flowers.

Q. 83. Which hormone acts as weed-killer?

Ans. 2, 4-Dichlorophenoxyacetic acid (2, Y-D).

Q. 84. What is abscission layer?

Ans. At the base of leaf and fruit, in most plant species, there comes a time during their life when they form a layer – due to this fruits and flowers are removed from the plant (abscission). This is called the Abscissien layer.

Q. 85. From where is Gibberellin derived firstly?

Ans. *Gibberella fujikuroi*, by Yabuta, 1935.

Q. 86. What are the major action on plants of GA?

Ans. (i) Elongation of internodes.

(ii) Bolting.

(iii) Dwarfism.

(iv) Parthenocarpy.

(v) Sex expression.

Q. 87. How does ABA regulate stomatal behaviour?

Ans. ABA plays an important role in the stress physiology of plants; under moisture stress condition the leaves accumulate much ABA. Under the high concentration of ABA, the ground cells lose H_2O and the stomata close. ABA, therefore, is involved in regulating stomatal aperture in water stressed plants. It, therefore, plays an important role in water conservation mechanism in plant.

Q. 88. How is apical dominance controlled by one endogenous hormone?

Ans. Auxin, particularly IAA, controls the apical meristematic growth in shoot apex.

Q. 89. What do you mean by bioassay of phytohormone?

Ans. The use of living material or system for determining the acitivity and concentration of hormones or enzymes is known as bioassay. When this bioassay is performed for plant hormones or phytohormones, it is known as Bioassay of Phytohormones.

Q. 90. Mention the precursor of biosynthesis of IAA.

Ans.

Tryptophan

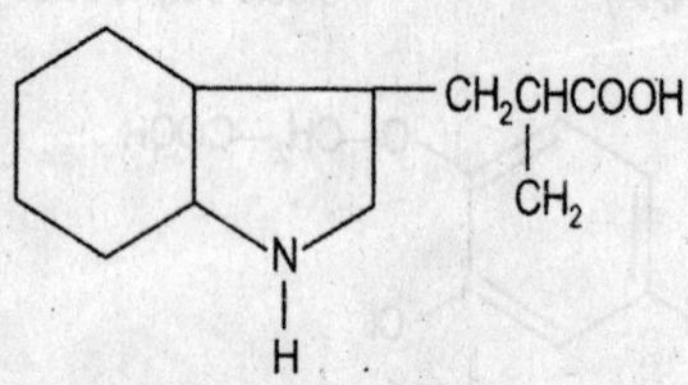

Q. 91. Name one inhibitor of photosynthesis.

Ans. D.C.M.U. ⇒ 3- (3, 4- Dichlorophenyl)-1, 1-dimethylurea.

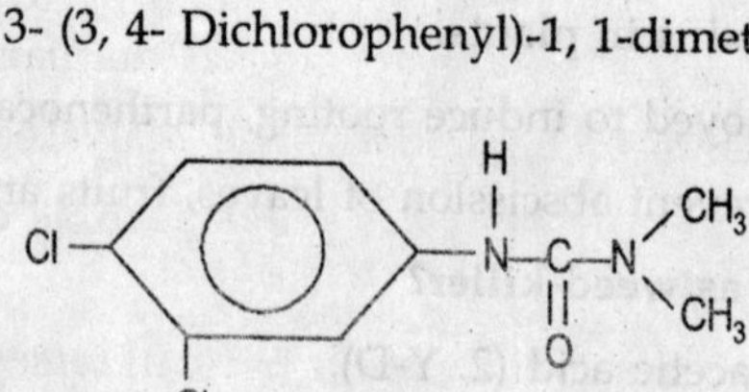

Q. 92. Name the plant hormones which play key roles in fruit ripening, internodal elongation, apical dominance and cell division.

Ans. (i) Fruit ripening ⟶ Ethylene

(ii) Internodal elongation ⟶ Gibberellin

(iii) Apical dominance ⟶ Auxin

(iv) Cell division ⟶ Cytokinin.

Q. 93. Name one antiauxin.

Ans. Napthylpthalamic acid (NPA).

Q. 94. How is a abscission beneficial to plants?

Ans. Abscission helps in the loss of worn-out leaves and fruits. During the dry season, the shedding of leaves prevents loss of excess water by transpiration and thus descication is prevented.

Q. 95. What is Bakanae disease?

Ans. Bakanae disease of rice plant is caused by the fungus *Gibberella fugikuroi*, it results in the increase in the length of internode, by which the length of the internode increase, the rice plants become very tall and ultimately breaks and dies.

Q. 96. What is primary dormancy?

Ans. Most seeds do not germinate immediately but rather spend a period when they germinate. This period of delay in initial seed germination is called primary dormancy.

Q. 97. What is stratification?

Ans. Temperate plants typically exhibit some type of deep dormancy, staying dormant until conditions are favourable for germination. Generally an accumulation over time of chilling is required to overcome deep dormancy. This chilling process is called stratification.

Q. 98. Name two chemicals which have inhibitory effects on germination.

Ans. Para-ascorbic acid and caumarins.

Q. 99. What is seed viability?

Ans. Seed viability is defined as the capacity of the seed to remain capable of germination for some specific period of time. The viability of different seeds may differ to great extent.

Q. 100. How is the viability of seeds is identified?

Ans. Viability of seeds can be tested with the help of Acetocarmine. If the embryo takes stains the seeds are viable and if not the seeds are not viable.

Q. 101. Name some families where seed coat imposes dormancy.

Ans. Plants belonging to families Leguminosae and Malvaceae.

Q. 102. If some seeds are treated by the juicy fruits such as oranges and tomatoes then what will happen?

Ans. The seeds do not germinate in such fruit's juice. In tomato fruits, ferulic acid is formed as growth-inhibitory substance.

Q. 103. What are the photoblastic seeds?

Ans. The germination of many seeds are affected by light – such seeds are called photoblastic seeds, *Epilobium hirsutum, Lactuca sativa* etc.

Q. 104. What is pre-dormancy?

Ans. It is the initial sub-phase of dormancy in which the organs may be induced to resume the normal course of growth by different treatments.

Q. 105. What is mid-dormancy?

Ans. From pre-dormancy, a plant organ may pass into full dormancy. At this stage it is difficult to induce the resumption of growth by making alteration in the environment.

Q. 106. What is post-dormancy?

Ans. In this subphase a plant organ gradually emerges out from the dormancy. At this stage it is easy to induce growth resumption.

Q. 107. Write an example where seed coat is impermeable to gases.

Ans. *Xanthium* sp.

Q. 108. Where are rudimentary embryos found?

Ans. Some members of Orobanchaceae, Orchidaceae, *Ginkgo biloba, Ranunculus* sp.

Q. 109. How is chemical scarification done?

Ans. Treating seeds with strong acids – such as H_2SO_4 – and washing in distilled water.

Q. 110. Give an example of a plant in which, dormancy of seeds can be broken by alternating temperature treatment?

Ans. *Poa pratensis.*

Q. 111. Which light stimulates germination and inhibits germination?

Ans. According to Flint & Mc.Alister, 1935-37, 37, the wavelengths between 520-700 nm (yellow, orange and red) stimulats germination and between 420-520 nm (blue -violet) inhibits it.

Q. 112. Which pigment is responsible for germination?

Ans. Phytochrome :

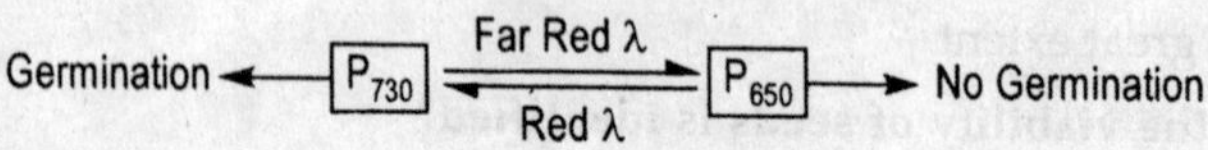

Q. 113. What is inhibitor-β?

Ans. It is present in the peel of dormant patoto tubers. The inhibitor β is responsible for checking the sprouting of buds located in the "eyes" of the potato tuber.

Q. 114. What is allelopathy?

Ans. Some seeds leach out chemicals which inhibit the germination of seeds of other plants. This phenomenon is called Allelopathy.

Q. 115. What do you mean by quiscence of dormancy?

Ans. In dormant plant organs, due to dormancy, there is arrest of growth or development. Because of non-availability of a desired environmental factor, it is referred to as quiscence of dormancy.

Q. 116. Write down the empirical formula of chlorophylla.

Ans. $C_{55}H_{72}O_5N_4Mg$.

Q. 117. Name one non-plastidial pigment.

Ans. β-cyanine.

Q. 118. Name one photorespiration inhibitor.

Ans. α-HPMS (α-hydroxy-2-pyridine-methanesulphonic acid).

Q. 119. What is innate dormancy?

Ans. Many seeds do not germinate under conditions that are usually favourable for germination. Such seeds are known as dormant seeds and the phenomenon is called innate dormancy.

Q. 120. What are induced dormancy?

Ans. Certain substances, such as ABA, induce dormancy in many seeds if applied exogenously. Due to this treatment by chemicals that induce dormancy, non-dormant seeds are changed into dormant seeds. Such type of dormancy is called imposed dormancy or induced dormancy.

Q. 121. How are seeds vernalised? or What do you mean by vernalisation?

Ans. Vernalisation can be defined as the method of inducing early flowering in plants by pretreatment of their seeds at very low temperatures.

Q. 122. How can you prove that a particular physiological process is controlled by the phytochrome system?

Ans. If the particular physiological process appears to be reverlable by the application of red and far red light, then it can be said that it is controlled by phytochrome.

Q. 123. What do you mean by florigen?

Ans. Florigen is the plant hormone responsible for the initiation of flowering in plants. It is synthesised in the older leaves and then transferred to the growing region where it initiates the floral bud initiation.

Q. 124. What is vernalin?

Ans. The effective substance that is formed during vernalisation (i.e. pretreatment of seeds at low temperatures to induce flowering) and the substance that can be transferred from one plant to another by grafting is called vernalin.

Q. 125. What are plastidial pigments?

Ans. The pigments localised in the chloroplast or plastids in plant cells are called plastidial pigments.

E.g. chlorophyll, caroteinoides.

Q. 126. What do you mean by pigment shift?

Ans. Under normal dry light condition, the phytochrome contains almost equal amount of P_{fr} and P_r but during the dark period the P_r declines, so that the P_r form is predominant. When the dark period is interrupted with a short period of red light then there is a rapid conversion of P_r back to P_{fr} which will decline again during further period of darkness. This phenomenon is called pigment shift.

Q. 127. How can plants be classified on the basis of their photoperiodic response?

Ans. According to photoperiodic response, plants can be classified into following types :

(i) *Long day (LD) plants* — Plants required a photo period of more than a critical length. E.g. Radish.

(ii) *Short day (SD) plants* — Plants requring a photo period of less than a certain critical length. E.g. Rice.

(iii) Day neutral plants — Plants whose flowering is not affected by the length of the day. E.g. Tomato.

(iv) Intermediate day plants — These plants flower when day length are neither too short nor too long. E.g. *Chenopodium.*

(v) Ambiphotoperiodic plants — The flowering in these plants, is quantitatively inhibited by intermediate day length. As for example : *Chenopodium rulenum.*

(vi) Amphophotoperiodic plants — In these plants flowering is delayed under intermediate day (12-14 hrs) but occurs rapidly under day lengths of 8 hrs or 18 hrs. E.g. *Madia elegans.*

Q. 128. What do you mean by delayed action of phytochrome?

Ans. Phytochrome action by different gene activation is called delayed action. E.g. Synthesis of anthocyanine by red light.

Q. 129. What is tanada effect?

Ans. Tanada effect is an expression of rapid action of phytochrome on membrane permeability. It is a non-photoperiodic response.

Q. 130. What do you mean by photoreversibility of phytochrome? Mention one phytochrome-centralled vital physiological process in plants.

Ans. Phytochrome exists in plant systems in 2 forms. P_r which absorbs red light (660 nm) and P_{fr} which absorbs far red light (730 nm). When red light exposure is given P_r is converted to P_{fr} and on giving for red light exposure P_{fr} is reversed to P_r form. This phenomena exhibited by phytochrome pigment is called photoreversibility.

One phytochrome controlled vital physiological process in plant is Flowering.

Q. 131. What is transpiration?

Ans. The excess of absorbed water is lost from the aerial parts of plants in the form of water vapours. This is called transpiration.

Q. 132. What is guttation?

Ans. In some plants such as garden nasturtium, tomato, *Calocasia* watery drops ooze out from the injured margins of the leaves where a main vein ends. This is called guttation.

It takes place usually early in the morning when the rate of water absorption and the root pressure are higher while the transpiration is very low.

Q. 133. What is hydathodes?

Ans. The phenomenon of guttation is associated with the presence of special types of stomata at the margins of the leaves which are called water stomata or hydathodes.

Q. 134. What is ascent of sap?

Ans. The upward movement of water through the stem (xylem) from the root to reach the topmost parts of the plants is called as ascent of sap.

Q. 135. What is the function of `nif' gene?

Ans. The nif gene is responsible for the regulation and synthesis of nitrogenase, needed for nitrogen fixation in plants.

Q. 136. Where does leghaemoglobin occur?

Ans. Leghaemoglobin occurs in the root nodules of leguminous plants.

Q. 137. Why is transpiration sometimes injurious for plants?

Ans. Transpiration is sometimes the cause of death of many plants when it exceeds the amount of water absorbed by the plant through the root. Also, it consumes energy and causes unnecessary absorption of excess water by roots.

Q. 138. What is the function of guttation?

Ans. Guttation in a way helps to release the high root pressure developed in the plant at times.

Q. 139. Define nif gene.

Ans. nif gene means nitrogen fixing gene. nif genes are 17 clusters of genes required for N_2 fixation.

Q. 140. Mention the role of the enzyme nitrogenase.

Ans. Enzyme nitrogenase catalyses the reaction of atmospheric nitrogen to ammonia (NH_3) and thus aids in nitrogen fixation by plants.

Q. 141. What are anti-transpirants?

Ans. Substances that retard or inhibit transpiration are called anti-transpirants.

E.g. Phenyl mercuric acetate (PMC), ABA.

FULL FORM

ATP	:	Adenosine triphosphate
ADP	:	Adenosine diphosphate
AMP	:	Adenosine monophosphate
GDP	:	Guanosine diphosphate
GTP	:	Guanosine triphosphate
NAD	:	Nicotinamide adenine dinucleotide
NADP	:	Nicotinamide adenine dinucleotide phosphate
FAD	:	Flavin adenine dinucleotide
FMN	:	Flavin mononucleotide
EMP pathway	:	Embden-Meyerhof-Parnas pathway
TCA cycle	:	Tricarboxylic acid cycle
Co-SH	:	Coenyzme-A
ETS	:	Electron transport system
Pi	:	Inorganic phosphate
RUDP	:	Ribulose 1, 5 diphosphate
PGAld	:	Phosphoglyceraldehyde
RUBISCO	:	Ribulose 1, 5 biphosphate carboxylase
PGA	:	Phospho glyceric acid
OAA	:	Oxalo acetic acid
PEPA	:	Phosphoenol pyruvic acid
ABA	:	Abscissic acid
2, 4-D	:	2, 4-dichlorophenoxyaceticacid
TIBA	:	Tri Iodo Benzoic Acid
CAM	:	Crassulacean Acid Metabolism
IPA	:	Indole Propionic Acid
DCMU	:	3- (3, 4-Dichlorophenyl)-1, 1-dimethylurea
α-HPMS	:	α-hydroxy-2-pyridine methanesulphonic acid.
2, 4, 5-T	:	2, 4, 5-Trichlorophenoxy acetic acid
IBA	:	Indole butyric acid
IAA	:	Indole acetic acid

NAMING/DISCOVERY/DISCOVERER

J. B. van Helmont (1648) showed increase in the dry weight of a willow plant due to anabolic process.

M. Malpighi (1671) indicated the preparation of food in the leaves.

J. Priestley (1772) proved that plants can purify the air.

J. Senebier (1782) showed that under illuminated condition, green plants have the capability of utilising CO_2.

N. T. de Sanssure (1804) first formulated the equation of photosynthesis

$$CO_2 + H_2O \xrightarrow[\text{Green plants}]{\text{Light}} (CH_2O)_n + O_2$$

Dutrochet (1837) proved the involvement of chlorophyll in photosynthesis.

J. V. Sachs (1862-64) pointed out starch as an end product of photosynthesis and considered chlorophyll containing cells as the site of photosynthesis.

Engleman (1882) showed that the blue-red region of visible spectrum is maximum involved photosynthesis.

Barnes (1898) coined the term photosynthesis and denoted the various components of the process.

F. F. Blackman (1905) Dark phase's reactions.

Van Neil (1931) demonstrated the photosynthetic process in bacteria.

Emerson and Arnold (1932) showed the existence of two pigment systems within light phase.

Robin Hill (1937) showed the photolysis of water during light phase.

Ruben & Kamen (1941) using O^{18} demonstrated that H_2O is the source of O_2 on photosynthesis.

Arnon Allen & Whatley (1954) utilised $C^{14}O_2$ to demonstrate fixation of CO_2 by isolated chloroplast.

M. Calvin, Bassham, Benson (1956) denoted the regeneration of Ribulose diphosphate in the C_3 cycle.

Hill and Bendal (1960) proposed the `Z' scheme for light reaction.

Park & Biggins (1964) proposed the term quantasome.

M. D. Hatch & C. Slack (1967) discovered the C_4 pathway in monocots.

Louis Pasteur (1857) described alcoholic fermentation in yeast.

G. Hevesy (1923) discovered the technique of tracing a biochemical pathway by using isotopes.

K. Lohmann (1929) discovered ATP, the source of ready energy in living cells.

Hans Krebs (1937) discovered TCA cycle.

F. Lipmann (1947) discovered coenzyme-A, a key compound in cell metabolism.

Embden, Meyerhof & Parnas discovered the glycolysis pathway.

Knoop (1905) proposed β-oxidation of fatty acids.

Decken & Tio (1959) first used the term photorespiration.

Kurosawa (1926) denoted gibberellin from rice plant infected with the fungus *Gibberella fugikuroi.*

T. Yabuta (1935) gave the name gibberellin.

Frits (1926) gave the name auxin.

Bothwick (1952) photoblastic effect on seed dormancy.

16 BIOCHEMISTRY

Q. 1. What is protein?

Ans. Polypeptidal chain of aminoacids is called protein which is an essential constituent of protoplasm.

Q. 2. What is chromoprotein?

Ans. In conjugative protein the nonprotein substance (prosthetic group) is a colouring matter. This is called chromoprotein.

Q. 3. What are isozymes?

Ans. Isozymes are enzymes which have multiple molecular forms in the same organism, catalysing the same reaction eg. lactic dehydrogenase (LDH), which has 5 isoenzyme forms.

Q. 4. Write down the structure of hexose sugar.

Ans.

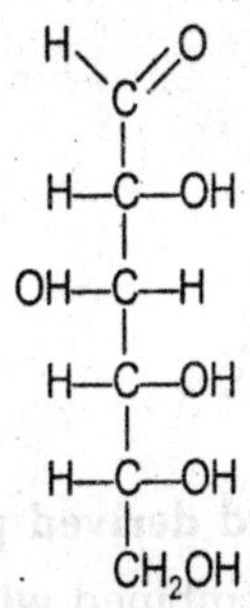

Q. 5. Distinguish between cofactor and prosthetic group.

Ans. Prosthetic group and cofactors are important part of the apoenzyme to become holoenzyme. But prosthetic group includes the organic part or protein part of the apoenzyme, whereas cofactor alway metalic part or non-protein part of the same.

Q. 6. What are oligosaccharides?

Ans. The simple compound carbohydrate contains a few monosaccharide units which are crystalline, water soluble and give sweet taste. These simple compounds are generally called oligosaccharides. E.g. Dextrose.

Q. 7. What are polysaccharides?

Ans. These compounds are composed of many monosaccharide units and so are called polysaccharides. E.g. Glycans, starch.

Q. 8. What is the function of coenzymes?

Ans. The coenzymes have an activating effect on the constitute enzymes.

Q. 9. Name two inhibitors of protein synthesis.

Ans. Puromycin ⟶ which causes the removal of an incomplete polypeptide chain from ribosome.

Chloramphenicol ⟶ which interferes in binding of ribosomes to m-RNA in microorganisms.

Q. 10. What are fats and oils?

Ans. Fats are esters of high molecular weight containing fatty acids and glycerol.

Fats which remain in the liquid form at room temperature are called oils.

Q. 11. Give a structure of a keto sugar.

Ans.

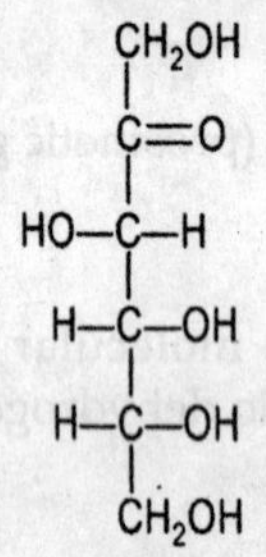

D-Fructose

Q. 12. What is the name of the simplest amino acid?

Ans. The simplest aminoacid is Alanine :

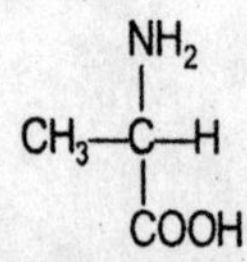

Q. 13. What are simple proteins, conjugated proteins and derived proteins?

Ans. Simple protein ⟶ These are pure proteins, uncombined with anything else.

Conjugative protein ⟶ In this types of protein, the molecules remain conjugated with another non-protein prosthetic group.

Derived protein ⟶ These proteins are not present in nature as such. They are found as product of hydrolysis of the native protein molecules.

Q. 14. What is metaprotein?

Ans. Metaproteins are coagulated by heat, soluble in excess of acid and alkali and are precipitated at isoelectric point or pH.

Q. 15. What are enzyme?

Ans. Enzymes are soluble organic catalysts manufactured by living cells but their activities do not depend upon the life of the cells.

Q. 16. What is apoenzyme, coenzyme and haloenzyme?

Ans. In conjugated enzyme the protein part is called apoenzyme and a prosthetic group is the coenzyme. Combination of apoenzyme with coenzyme constitutes the haloenzyme.

Q. 17. What is enthalpy (H)?

Ans. The heat content of a system is called enthalpy (H).

Q. 18. What is heme protein?

Ans. A protein containing a heme as a prosthetic group is called heme protein.

Q. 19. What is kerotins?

Ans. Kerotins are insoluble protective or structural proteins consisting of parallel polypeptide chains in α-helical or β conformatins.

Q. 20. What is zinc finger?

Ans. It is a specialized protein malif invalued in DNA by some DNA-binding proteins. It is characterized by a single atom of zinc coordinated by four cys residues.

17 CELL BIOLOGY & GENETICS

Q. 1. What is resolution?

Ans. The smallest distance that may seperate two points on an object and still permit their observation as distinct seperate points; resolution determines whether small objects that are close together can be distinguished as separate objects.

$$\pi = 0.61\ \lambda/\text{N.A.}$$

π = Smallest distance between 2 points resolved.

λ = Wave length of light used

N.A. = Numerical aperture.

Q. 2. What is numerical aperture?

Ans. The quantity n sin α is called the numerical aperture of the objective lens abbreviation NA.

λ = Wave length of light used

n = Refractive index of the medium,

α = Angle which marginal rays form with axis.

Q. 3. Why use the immersion oil in microscope under high power?

Ans. By increasing the numerical aperture value some microscope lenses are designed to be used with a layer of immersion oil between the lens and the specimen. Immersion oil has a higher refractive index than air; therefore, it allows the lens to receive more of the light transmitted through the specimen.

Since the refractive index of immersion oil is about 1.5, the maximum numerical aperture for an oil immersion lens is about $1.5 \times 0.94 = 1.4$. The rosution of an oil immersion lens is, therefore, about 0.2 μm which can be calculated following :

$$\pi = 0.61\ \lambda/\text{NA}$$

$$= 0.61 \times 450\ /\ 1.4 = 196\ \text{nm} \approx 0.2\ \pi\text{m}$$

Q. 4. Who is the father of microscope?

Ans. Antoni van Leeuwenhoek.

Q. 5. Write down the functions of different light microscopes.

Ans. Bright field (unstained) : Passes light directly through specimen; unless cell is naurally pigmented or artificially stained, image has little contrast.

Bright field (stained) : Straining with various dyes enhances contrast, but most staining produces require that cells be fixed (preserved).

Phase contrast : Enhances contrast in unstrained cells by amplyfying variations in density within specimen; especially useful for examining living unpigmented cells.

Q. 6. What is the special character of fluorescent microscopy?

Ans. Fluorescent microscopy invalues measurements based on the absorption of light by a specific fluorescent probe. E.g. Mushroom toxin, phallodin.

Q. 7. Who discovered microscope?

Ans. Francis Janssen and Zacharias Janssen, the two sepectacle sellers of Holland discovered microscope in 1590.

Q. 8. Who discovered electron microscope?

Ans. Electron microscope was discovered by Knoll and Ruska in 1931.

Q. 9. What is the resolving power of (i) human eye, (ii) light refracting microscope (iii) electron mocroscope?

Ans. (i) 0.1 mm (or 100 μm), (ii) 0.2 μm (or 200 nm), (iii) 0.5 nm

Q. 10. Who first discovered the cell?

Ans. Robert Hooke, a British scientist, in 1665.

Q. 11. What is plasma-membrane?

Ans. The thin, elastic and semipermeable living membrane that surrounds the protoplasm of a cell is called plasma membrane or plasmalemma or cell membrane.

Q. 12. What is phagocytosis?

Ans. Phagocytosis means "eating of cells" i.e. active engulfing of a solid particle into the cell by invagination of the cell membrane. Thus, the particle is incorparated within the cell in a membrane-bound vesicular form known as phagocytic vacuole or phagosome.

Q. 13. What is pinocytosis?

Ans. Pinocytosis means "drinking of cells" i.e. active intake of a fluid. The cell membrane invaginates to form a narrow channel, the pinocytatic channel, within which the fluid enters. Then, from the inner end of the channel, small fluid-filled vesicles are pinched-off one by one into the cytoplasm.

Q. 14. What is reverse pinocytosis?

Ans. Reverse pinocytosis is opposite of pinocytosis or "vomiting of cells". It is also called 'emeiocytosis'. For this the membrane covering of a vesicle fuses with the cell membrane to extrude the contents of the vesicle leaving the cell membrane intact.

Q. 15. What is cell wall?

Ans. Cell wall is the semi-rigid, laminated, non-living cellulose covering present outside the cell membrane of all plant cells and bacteria.

Q. 16. Who first showed the presence of mitochondria in living cells?

Ans. Kolliker (1850) first showed the presence of mitochondria in living cells (muscle cells).

Q. 17. What is the function of mitochondria?

Ans. Mitochondria helps in aerobic cellular respiration by carrying out TCA cycle, β-oxidation of fatty acids, electron transport and oxidative phosphorylation.

Q. 18. What is rough ER? Why is it so named?

Ans. The endoplasmic reticulum (ER) containing ribosomes on their surface is called rough ER. It is made up of a network of inter-communicating cisternae having ribosomes on their surface.

It is so named because its surface appears to be rough and granular due to the presence of ribosomes.

Q. 19. In which type of cells rough surfaced endoplasmic reticulum is found?

Ans. Rough endoplasmic reticulum is found in those cells which are actively synthesising proteins, e.g. cells of liver, pancreas.

Q. 20. Name the cells where ER is absent.

Ans. Endoplasmic reticulum is absent in prokaryotic cells, cells of fungi and lower plant and animal cells like ovum, mature mammalian RBC.

Q. 21. What are the functions of E.R.?

Ans. The functions are :

(i) Forms a fine network in cytoplasm that gives mechanical support to their cytoplasmic matrix.

(ii) Helps in formation of nuclear membrane, golgi body and cell wall.

(iii) Provides large surface area of membrane for exchange of materials between cytoplasm and extracellular fluid.

(iv) Helps in synthesis of proteins, lipids, steriods and lipoproteins.

(v) Concerned with glycogenolysis and detoxification.

Q. 22. What is ribosome?

Ans. Ribosomes are small, dense, granular, non-membranous organelles that are made up of r-RNA, ribonucleoprotein and concerned with protein synthesis.

Q. 23. In 70s ribosomes, what does the letter s stand for?

Ans. 's' here stands for sedimentation coefficient or Swedberg units.

Q. 24. Mention one major role of golgi apparatus.

Ans. The Golgi apparatus is mainly concerned with the formation and packaging of materials for export (secretion) from the cell across the cell membrane by the process of reverse pinocytosis.

Q. 25. Name one plant cell and one animal cell where nucleus is absent.

Ans. Plant cell → sieve tube

Animal cell → matured mammalian RBC.

Q. 26. Which number of human chromosomes have nuclear organiser?

Ans. There are 6 chromosomes – 13, 14, 15, 21, 22 and Y.

Q. 27. What is the full form of SAT?

Ans. Sine Acido Thymonucleinico, without thymonuclic acid.

Q. 28. What is L chromosome?

Ans. Limited chromosomes or L chromosomes are so called because they are large and limited to the germ line. Found in the family seiaridae (Dipthea : insecta).

Q. 29. What is m chromosome?

Ans. m chromosome or minute chromosomes are so called because of their extremely small size (0.5 μ or less) eg. different species of bryophytes, coreidae (heteroptera).

Q. 30. What do you mean mega chromosome?

Ans. Certain large chromosomes, up to 15 times the length of normal chromosomes, have been described in a few species of hybrids.

Q. 31. Who first proposed the term "chromosome"?

Ans. Waldeyer in 1888.

Q. 32. Where do you find the lowest number of chromosomes in plants?

Ans. Somatic cell of *Happlopappus gracilis,* family Compositae.

Q. 33. How is the chromosome size affected artificially?

Ans. Chromosome size is markedly affected by different kinds of chemicals eg colchicine, 8-hydroxyquinoline etc.

Q. 34. What is chromatid?

Ans. Each metaphase chromosome appears to be longitudinally divided into two identical parts each of which is known as chromatid.

Q. 35. What is diffuse centromeres?

Ans. In many insects eg. most homopteran and hemipteran insects, the centromeric activity is non-localised and spread over the entire chromosome length, such centomeres are known diffuse centromeres.

Q. 36. What do you mean by NOR?

Ans. Secondary censtrictions are called nucleolus organiser regions (NOR) i e. site of nucleolus organigation.

Q. 37. What is NOC?

Ans. Satellite are referred to as nucleolus organiser chromosomes.

Q. 38. What is sat DNA?

Ans. Sat DNA or satellite DNA contains short repetitive DNA sequence; it is constitutive heterochromatins. It is found in centromeric regions of all chromosomes.

Q. 39. What is chromomeres?

Ans. In some species, maize, amphibia etc., chromosome during the first prophase of meiosis, i.e. pachytene, show small bead-like structures called centromeres. It is made up of DNA and proteins.

Q. 40. What is chromatin?

Ans. The nucleoplasm of eukaryotic nucleus contains many thread-like coils. These thread-like structures are known as the chromatin or chromatin fibres or nucleoprotein fibres.

Q. 41. What is euchromatin?

Ans. The structural or functional DNA, i.e. gene-bearing portion of chromatin, is called euchromatin or active chromatin. Euchromatin always takes less stains because it remain less condense and less compact during all stages of cell cycle.

Q. 42. What is heterochromatin?

Ans. Certain regions of the chromatin appear to be more densely compact than others during all stages of cell cycle and stained more densely by fuelgen stain or acetoorcein or acetocarmine and said to contain heterochromatin. These regions generally nonfunctional regions.

Q. 43. What is constitutive heterochromatin?

Ans. The heterochromatin which remains cendensed, i.e. transcriptionally silent in all cells at all times of the cell cycle and are often found near the centromere, telomere, NOR, Satellite and intercalary segment of chromosome.

Q. 44. What is facultative heterochromatin?

Ans. The heterochromatin which is not permanently maintained in the condense stage is called facultative heterochromatin. For example, x chromosome in mammalian female.

Q. 45. What is unique DNA?

Ans. Unique DNA consists of those DNA segments where base sequences are present only in a single copy per genome. Eg. genes for histones, ribosomal RNA, storage proteins etc.

Q. 46. Which eukaryatic organism has lowest chromosome number?

Ans. *Ascaris megalocephala* has two chromosomes per somatic cell.

Q. 47. Which organism has highest chromosome number?

Ans. *Ophioglossum reticulatum* has 1,260 chromosomes per somatic cell.

Q. 48. What is central dogma?

Ans. F.H.C. Crick's idea that genetic information flows from DNA to RNA & RNA to protein is called central dogma.

Q. 49. What is Karyotype?

Ans. The general morphology of a set of chromosomes at the metaphase stage of an individual or species is known as karyotype. The karyotype was first defined in 1926 by Delaunay.

Q. 50. What is Kinetochore?

Ans. In the primary constriction region, the plate-like pratoplasmic body connecting the centromere and spindle fibra is known as kinetochore. This term was referred by Schrader.

Q. 51. What are introns and exons?

Ans. Introns : Those regions of a DNA which do not form part of mRNA and are removed during mRNA formations. These are noncoding sequenced.

Exons : These regions of a DNA contain gene which form part of mRNA and code for amino acids of proteins.

Q. 52. What do you mean by extrachromosomal inheritance?

Ans. These are due to the extrachromosomal DNA. In eukaryotes mitochondrial or chloroplast DNA are responsible for such inheritance. The transfer of DNA information do not maintain Mendelian Inheritance e.g. male sterility in plant.

Q. 53. Cite the differences between B and Z form of DNA.

Ans.

B DNA	*Z DNA*
1. B form has right-handed helical sense.	1. Z form has left-handed helical sense.
2. One complete helix is 34A°.	2. One complete helix is 45A°.
3. Glycosidic torsion angle is anti in BDNA.	3. Clycosidic torsion angle is syn in ZDNA.

Q. 54. What is nucleosome?

Ans. A bread-like structure of chromatin fibres, each composed of a histone octamer and 146 bp of DNA.

Q. 55. What is codon?

Ans. Codon is a sequence of 3-nucleotide bases present in an m-RNA and codes for a particular amino acid or stop signal of protein synthesis.

Q. 56. What is the full form of V.N.T.R.?

Ans. Variable Nucleotide Tandem repeat.

Q. 57. What is the C-value DNA?

Ans. The haploid DNA content in an individual is described as the C-value of DNA.

Q. 58. What is the AC-DS system?

Ans. Activator-dissociation system is the most popular system of maize controlling elements i.e. transposomes, and was discovered by Barbara Mc.clintock in 1950.

Q. 59. What is B-form DNA?

Ans. The normal structure of DNA as commonly found in living system is called B-form DNA. It follows the right-handed double helix structure of DNA.

Q. 60. What is endomitosis?

Ans. 2 or more chromosomes replication not followed by nuclear or cell division leads to endopolyploidy, i.e. each replication doubles the existing chromosome number of the cell.

Q. 61. What is genome?

Ans. The haploid set of chromosomes of an organism is called Genome. All the members of a genome are distinct from each other in gene content and often in morphology.

Q. 62. What is cistron?

Ans. A segment of DNA which codes for one polypeptide. It is generally accepted as a synonym for gene.

Q. 63. What is bivalent?

Ans. The structure formed by pairing of two homologous chromosomes during meiosis, seen during pachytene and onward till the beginning of anaphase I.

Q. 64. What is asynapsis?

Ans. The failure – partial or complete – of pairing between homologous chromosomes during meiosis.

Q. 65. What do you mean by anticodon?

Ans. The three bases of a t-RNA that are complimentary to the three bases of mRNA coding for an amino acid.

Q. 66. What is structural gene?

Ans. A gene which codes for the amino acid sequence of a polypeptide.

Q. 67. What is sex chromosome?

Ans. A chromosome which differs in number or kind between the male and female individuals of the same species and is invalued in sex determination.

Q. 68. What is pseudogene?

Ans. An inactive gene similar to a normal gene – occuring as a normal component of an eukaryotic genome, may arise due to mutation following duplication of the concerned gene.

Q. 69. What is lampbrush chromosomes?

Ans. In oocytes of fishes, amphibia, reptiles and birds there is present lampbrush-like appearance due to the formation of lateral loops of DNA called lampbrush chromosomes. Flemming discovered the lampbrush chromosomes in 1882.

Q. 70. Distinguish between centrosome and centromere.

Ans.

Centrosome	*Centromere*
1. It is a cytoplasmic organelle of animal cell and lower plant cell.	1. It is a part of chromosome visible during cell division.
2. Made up of lipo-protein.	2. Made up of nucleoprotein.
3. It consist of the centriole which divide at onset of cell-division.	3. It consists of 4 centromeric chromosomes dividing at the time of anaphasic separation of chromosomes.
4. It forms the spindle fibre in animal cell and helps in the movement of lower plant cell.	4. It helps in the attachment of spindle fibres to the chrosmosomes during anaphase.

Q. 71. What do you mean by polytene chromosome?

Ans. E. C. Balbiani (1881) discovered a special kind of giant chromosome in the tissues of salivary gland, trachea, Malpighian tubules of Diptera larva (insects), during inter-phase called Polytene chromosome. The name was given by Kollar because of the presence of multiple chromonemata.

Q. 72. What is DNA finger-printing?

Ans. The process of identifying a person by studying RFLPs (Restriction Fragment Length Polymorphism) of his or her genome is known as DNA finger printing.

Q. 73. What is halandric gene?

Ans. The genes present in the Y chromosome, helping in paternal transmission of characters are called Halandric gene.

Q. 74. What is jumping gene?

Ans. The functional protein of DNA, i.e. gene sequences, which can change positions within one chromosome or between chromosomes is known as transposons or jumping genes.

Q. 75. What is oncogene?

Ans. The genes responsible for neoplastic growth or cell proliferation in organisms are called oncogene.

Q. 76. What is overlapping gene?

Ans. The genes which are functional within large-sized genes – as in case of some viral (e.g. ΦX174) genes – is called overlapping genes.

Q. 77. What is plasma gene?

Ans. The extra-nuclear inheritance brought upon by the DNA present in the cytoplasmic organelles is called plasma gene.

Q. 78. What is pseudo gene?

Ans. The part of the DNA not involved in transcription is known as pseudo gene.

Q. 79. What is split gene?

Ans. The entire DNA can be divided into functional sequences called exon and non-functional repetitive sequences called introns. So the functional sequence (gene) is splitted by non-functional sequence and is called split gene is the characteristics of eukaryotic DNA.

Q. 80. What is muton?

Ans. The smallest sequence of DNA capable of undergoing mutation which should have at least a pair of nitrogen bases.

Q. 81. What do you mean by recon?

Ans. The smallest DNA segment which includes crossing over is called recon.

Q. 82. What is complon?

Ans. The particular DNA segment coding for more than one polypeptide is called complon.

Q. 83. What is replicon?

Ans. The DNA segment consisting of more than one complon is called replicon.

Q. 84. What is gene?

Ans. Gene is a functional fragment of DNA nucleotide of synthesising a specific protein and acting as the unit of inheritance.

Q. 85. How would you classify DNA on the basis of organisation?

Ans.

Type	Nature	Helical Diameter	Base pairs per turn of 360⁰	Axial Rise (n)
A	Right-handed	23A⁰	11	2.56A⁰
B	Right-handed	19A⁰	10	3.32A⁰
C	Right-handed	19A⁰	9.33	3.32A⁰
D	Right-handed	20A⁰	8	3.03A⁰
Z	Right-handed	18A⁰	12	3.70A⁰

Q. 86. What is centromeric index?

Ans. Study of the position of the centromere on the different chromosomes of a particular species i.e. it is the ratio of short arm length to that of total chromosome length.

Q. 87. What is idiogram?

Ans. When the chromosomes are represented diagrammatically and arranged in homologous pairs in the order of decreasing length.

Q. 88. What is selfish DNA?

Ans. The repetitive sequences of eukaryotic DNAs which do not take part in protein synthesis.

Q. 89. What is Lyon's hypothesis?

Ans. The second X chromosome in human female become heterochromatinised and forms the Barr body to prevent the double dosage of x chromosomes. It is also called dosage compensation.

Q. 90. What is supernatant RNA?

Ans. The other name for t-RNA.

Q. 91. What are nonsense codons?

Ans. The codons like UAA, UAG and UGA, which do not code for any amino acid.

Q. 92. What is initiating codon?

Ans. The codon AUG coding for the amino acid methionine initiates the process of polypeptide synthesis and is known as initiating codon.

Q. 93. What is okazaki fragments?

Ans. The small fragments of DNA formed after the RNA primer is attached the lagging strand during replication of DNA are known as Okazaki fragments.

Q. 94. What is barr body?

Ans. The inactive second x chromosome in human female denoted by Murray Barr (1949). It is to prevent the double dosage of x chromosome in female (Dosage compensation of Lyon). It is used to test the sex of individual and super female syndrome.

Q. 95. What is G_0 stage?

Ans. G_0 stage is the arrested G_1 stage in a cell cycle. If a cell enters in cell cycle at the time the extracellular signalling is not proper then the cell enters in a bipassic phase called G_0 phase where cells may stay for 1 hour, 2 hours, more or till organism death.

Q. 96. What is D-phase?

Ans. The cytokinesis in a cell devision is referred to as D phase.

Q. 97. What is ribozyme?

Ans. The enzyme – like activity of ribosome dissolving the nucleus at the end of prophase is called ribozyme.

Q. 98. What is S.P.F.?

Ans. The S-phase promoting factor responsible for activating the process of DNA replication during S phase of cell cycle.

Q. 99. What do you mean by nucleotide and nucleoside?

Ans. In DNA there is present sugar deoxyribose, a phosphate group and a nitrogenous base which may be purine or pyrimidine known as nucleotide.

Deoxyribose and nitrogenous base together form nucleoside of a DNA.

Q. 100. What is histone protein?

Ans. Histones are small proteins which are basic because they have a high content of amino acids, arginine and lysine. Being basic, histones bind tightly to DNA which is acid. The 4 main histones I L_2A, H_2B, H_3 & H_4 are very similar in different species and present in equimolar amounts. There is also another histone – H_1.

Q. 101. What is point mutation?

Ans. A mutation may arise due to a change in a single base pair of a gene, such mutations are called point mutation.

Q. 102. What is gene conversion?

Ans. A process of non-reciprocal recombination in which an allele is present in an excess of the other in the recombination product.

Q. 103. What is allele or allelomorph?

Ans. A pair of genes that represents contrasting alternatives of a characteristics and located at the same locus in the homologous chromosomes is known as allele or allelomorph.

Q. 104. What is a test cross?

Ans. Test cross is a cross bestween heterozygous F_1 hybrid and the double recessive homozygous parent. This is used to determine whether the individuals exhibiting dominant characteristics are homozygous or heterozygous.

Q. 105. What is epistasis?

Ans. When two independent genes affect the same characteristic of an organism and expression of one gene hides the effect of another gene then this phenomenon is known as epistasis.

Q. 106. What are complementary genes?

Ans. In this case, each pair of alleles is located in separated pair of chromosomes. When both the alleles are essential for the expression of a particular characteristic such genes are known as complementary genes.

Q. 107. What is supplementary genes?

Ans. When two independent but dominant genes are interacting in such a way that a different type of phenotype expression is developed, such genes are known as supplementary genes.

Q. 108. What is multiple allele?

Ans. When two or more alleles control one character of an individual then these genes are collectively called multiple alleles.

E.g. ABO blood group is controlled by more than one allele.

Q. 109. What is synaptonemal complex?

Ans. A protenaceous structure about $1000A^0$ wide formed between homologous chromosomes during synapsis of zygotene, meiosis, considered to facilitate recombination or crossing over.

Q. 110. What is Wobble base pair?

Ans. According to wobble hypothesis only the first two positions of a triplet codon on mRNA have a precise pairing with the bases of the tRNA anticodon. Such interaction between third bases is referred as wobble base pair.

Naming/Discovery/Discoverer

1856-1863	Gregor Johann Mendel	Conducted his famous pea experiments concerning gene segregation.
1859	Charles Darwin	Published on the origin of species; modern theory of evolution.
1871	Fredrich Miescher	Isolated DNA
1882-1885	E. Strasburger, Walther Flemming	Showed the nucleus contained chromosomes.
1902	Archibald Garrod	Identified the first human genetic disease.
1902	Walter Sutton, Theodor Boveri	Proposed the chromosome theory of heredity.
1905	William Bateson	Called the science of heredity "genetics".
1905	W. Bateson, R. C. Punnett	Demonstated linkage between genes.
1909	W. Johannsen	Introduced the word "gene".

1931	Harriet Creighton, Barbara Mc. Clintock	Showed genetic recombination in maize.
1953	James Watson, Francis Crick	Proposed double helical model for DNA.
1958	Matthew Meselson, Franklin Stahl	Proved the semiconservative model for DNA replication.
1961	Francois Jacob, Jacques Monod	Put forward the operon model for the regulation of gene expression in bacteria.
1966	Marshall Nirenberg, H. Gobind Khorana	Worked out the complete genetic code.
1977	Walter Gilbert & Frederick Sanger	Devised methods for sequencing DNA.
1977	Phillip Sharp and others	Discovered introns in eukaryotic genes.

Inheritable Diseases

1. **Albinism** – Failure of the tyrosine system to oxidise tyrosine to melanin, so that skin is fair, hair white and eye pink. It is due to a recessive autosomal gene.
2. **Alkaptonuria** – Presence of alkaptone in urine, resulting from only partial oxidation of phenylamine and tyrosine. Caused due to dominant autosomal gene.
3. **Brachydactylia** – Having very short fingers. It is due to dominant homozygous abnormal autosomal genes.
4. **Christmas disease** – Also known as Haemophilia B, caused by hereditary deficiency of clotting Factor IX (Plasma Thromboplastin Compunt – PTC) X - linked 4 recessive.
5. **Colour Blindness or Achromatopsia** – Due to X-linked recessive gene, cause complete colour blindness as only monochromatic grey is visible.
6. **Congential deafness** – Deafness existing from birth due to recessive autosomal gene.
7. **Cystinuria** – A metabolic disorder in which cystine appears in the urine. It is due to a recessive autosomal gene.
8. **Epilepsy** – Results from disordered electrical acivity of brain – causes unconsciousness and convulsions. It is due to recessive autosomal gene.
9. **Glactosemia** – Excess of galactose in the blood. It is the result of congenital deficiency and one cause of mental subnormality. It is due to recessive autosomal gene.
10. **Hare lip** – A congential defect in the lip, a fissure extending from the margin of the lip to the nostril. A dominant autosomal disorder.
11. **Migrain** – Periodic throbbing headache caused due to a dominant autosomal gene.
12. **Phenylketonuria** – Metabolites of phenylanine the phenylketones present in the urine. An autosomal recessive disease resulting in the mental subnormality.

13. **Schizophrenia** – A mental illness characterised by disorganisation of patient's personality causing lifelong ill health. A recessive autosomal disease.
14. **Sickle cell anemia** – A hereditary haemolytic anemia : The red cells are crescent-shaped. A recessive autosomal peculiar to negroes.

Anomalies due to Chromosomal Abrerration in Human

1. **Klinefelter's syndrome** – (XXX-47)-2n or 44 + XXY = 47 Trisomic (Aneuploidy). Individuals are males despite the presence of two X (chromosomes external genitalia normal, sperms not produced. Breast development common, mentally retarded).
2. **Edward's syndrome** – (Trisomy-18) 2n + 1 = 47 Trisomic (Aneuploidy). This is due to addition of extra 18th chromosome i.e. 2n + 1 = 47, more common in females. Many malformations, mental retardation, death occurs generally at 3-4 months of age.
3. **Turner's syndrome** – 45X (AAXO) Monosomic i.e. 44 + X = 45 or (2n - 1). They have only one X chromosome, sterile females, short, mentally retarded, breast absent, genitalia infantile.
4. **Down's syndrome** or **Mongolism** – Due to addition in the 21st chromsome (2n = 47). Individuals have short height, are mentally retarded, males sterile, females bear children.
5. **Patau's syndrome** – (Trisomy-13), Trisomic (Aneuploidy) 13th pair of autosomal chromosomes belonging to D-group. Newborns with multiple malformations. Mentally retarded, dies within hours or days.
6. **Myelogenous Leukemia** – A malignant proliferation of the leucopoetic tissues in bone marrow producing an abnormal increase in leucocytes in the blood with immature cells among them. It is due to delation in 22nd chromsome.
7. **Cri-du-chat syndrome** – Produced by the partial loss of one of the number 5-chromosomes leading to mental subnormality. Characterised by toneless cat-like cry in infancy.

18 BIOTECHNOLOGY

Q. 1. What do you mean by androgenesis?

Ans. Development of plants from male gametophyte is known as androgenesis.

Q. 2. Define antibiotic.

Ans. Antibiotic is a chemical substances, produced by microorganisms and, synthetically, that has the capacity to inhibit the growth of an event to destroy, bacteria and other microorganisms in dilute solution.

Q. 3. What is attenuated vaccine?

Ans. When the whole pathogenic organism is treated with chemical, radiation, or other means to render it incapable of producing infection then it is called attenuated vaccine.

Q. 4. What is autoradiograph?

Ans. Photograph showing location of radioactive substances in cells or tissue, obtained by exposing photographic emulsion in dark to radioactive emissions from the preparation known as autoradiograph.

Q. 5. What do you mean by batch culture?

Ans. A suspension culture in which cells grow in a finite volume of nutrient medium and follow a sigmoid pattern of growth called batch culture.

Q. 6. What is base sequence?

Ans. The specific order of purine and pyrimidine bases in a polyneucleotide chain called is base sequence.

Q. 7. Define bioassay.

Ans. Bioassay is the method of measuring the concentration or activity of a substance by determining its effect on a living organism.

Q. 8. What do you mean by clone?

Ans. When an individual is obtained from a single organism through asexual reproduction then it is known as clone.

Q. 9. Define complementary DNA.

Ans. DNA which are synthesised by reverse transcriptase from an RNA template are called complementary DNA.

Q. 10. What is cosmid?

Ans. A vector for carrying large DNA fragment into host cell.

Q. 11. What is cybrid?

Ans. The plant or cell which is a cytoplasmic hybrid produced by fusion of a protoplast and cytoplast.

Q. 12. What do you mean by denaturation?

Ans. Alternation of material, usually protein, through heating or chemical treatment, as it loses its characteristic activity.

Q. 13. What is differentiation?

Ans. Modification in structure and function of the parts of an organism, owing to increase in specialisation.

Q. 14. Define DNA probe.

Ans. A short, labelled DNA segment complementary in base sequence to part of another DNA, which is used to identify or isolate the particular DNA from a mixture through its ability to bind specifically with the target DNA.

Q. 15. What is electrophoresis?

Ans. Separation of molecules based on their differing electrical charges is known as electrophoresis.

Q. 16. Define embryogenesis.

Ans. A pathway of differentiation which is characterised by the formation of organised structures that resemble zygotic embryos is called embryogenesis.

Q. 17. What do you mean by embryoids?

Ans. Embryoids are embryo-like structures produced as a consequence of differentiation process such as embryogenesis or androgenesis.

Q. 18. What do you mean by endonuclease?

Ans. It is an enzyme that hydrolyses DNA by breaking bonds in the middle of the chain.

Q. 19. What is explant?

Ans. A plant organ or piece of tissue used to initiate a tissue culture known as explant.

Q. 20. What is genetic engineering?

Ans. The experimental manipulation of DNA (or RNA) of different species producing recombinant DNA, which include some gene from both species.

Q. 21. Define germplasm.

Ans. The total genetic variability available to species is called as germplasm.

Q. 22. What do you mean by genetic information?

Ans. The information contained in a sequence of nucleotide base in a DNA (or RNA) molecule.

Q. 23. What do you mean by vaccine?

Ans. A preparation of pathogenic microorganism which is treated to destroy virulence when introduced into the body, causing the body to produce antibodies and so conferring immunity.

Q. 24. What do you mean by transgenic organism?

Ans. A generally manipulated organism containing in its genome one or more inserted genes of another species.

Q. 25. What is Ti-plasmid?

Ans. Ti-plasmid is a plasmid from *Agrobacterium tumefaciens*, used as a plant vector.

Q. 26. Define synthetic seed.

Ans. Embryoid preserved in a thin layer of calcium alginate is known as synthetic seed.

Q. 27. What is recombinant DNA?

Ans. Hybrid DNA produced by joining pieces of DNA from different organisms together in vitro.

Q. 28. Define restriction endonuclease.

Ans. Enzyme that cleaves DNA at specific nucleotide sequence, often remain single stranded region known as restriction endonucleoase.

Q. 29. What do you mean by restriction enzyme?

Ans. Enzyme that cuts DNA at specific DNA sequence.

Q. 30. What is reverse transcriptase?

Ans. Enzyme that polymerises a molecule of DNA using RNA as the template for determining sequence.

Q. 31. What is SCP?

Ans. SCP is single cell protein. Cell or protein extracts of microorganism are grown in large quantities for use as human or animal protein supplements.

Q. 32. What do you mean by gene library?

Ans. Gene library is a collection of DNA fragments (cDNA genomic DNA) which have been cloned in a vector.

Q. 33. Define organogenesis.

Ans. Organogenesis is a type of morphogenesis which results in the formation of roots, shoots and/or floral organ.

Q. 34. What is somatic embryogenesis?

Ans. Development of embryos from somatic cells in culture whose structure is alike to zygotic embryos found in seed and with analogous embryonic organs such as cotyledons or leaves.

Q. 35. What do you mean by operon?

Ans. Operon is a genetic unit consisting of adjacent genes that function coordinately under the joint control of an operator and a repressor.

Q. 36. What is myeloma?

Ans. Myeloma is a antibody producing tumour cells.

Q. 37. What is immunisation?

Ans. Administration of antigen to animal or organism to stimulate production of antibodies by that organism; also administration of antigen, antibodies or lymphocytes to organism to produce corresponding active, passive or adaptive immunity is known as immunisation.

Q. 38. What is nick translation?

Ans. Nick translation is a procedure for radio labelling DNA in vitro.

Q. 39. Define morphogenesis.

Ans. Morphogenesis is a developmental pathway in differentiation which result in the formation of recognised tissue.

Q. 40. What do you mean by monoclonal antibodies?

Ans. Those antibodies which are derived from a single source clone of cell and which recognise only one kind of antigen.

Q. 41. What is interferon?

Ans. Interferon is a glycoprotein, important in immune function and thought to inhibit viral infection.

Q. 42. What is gameto-clonal variation?

Ans. When variation is observed among plants regenerated from gametic culture it is called gametoclonal variation.

Q. 43. What is protoclonal variation?

Ans. When variation is observed among plants regenerated from protoplast culture it is called protoclonal variation.

Q. 44. What do you mean by somatic hybridisation?

Ans. A technique of fusing protoplast from two contrasting genotypes for producing of hybrid or cybrid which contains various mixtures of nuclear and/or cytoplasmic genotype.

Q. 45. What is transduction?

Ans. The transfer of bacterial genes from one bacterium to another by a bacteriophage particle.

Q. 46. Define bioreactor.

Ans. Bioreactor is a apparatus used for bioprocessing.

Q. 47. What is concatamer?

Ans. When DNA is made up of repeated unit length DNAs then it is known as concatamer.

Q. 48. What do you mean by cytoplast?

Ans. The enucleated protoplast is known as cytoplast.

Q. 49. What is attenuator?

Ans. A nucleotide sequence in the 5′ region of a prokaryatic gene (or in its RNA) that cause premature termination of transcription, possible by forming a secondary structure is called attenuator.

Q. 50. What is hybridoma?

Ans. A method of creating pure and uniform antibodies against a specific antigen.

Q. 51. What is somaclonal variation?

Ans. When variation is observed among plants regenerated from somatic culture.

Q. 52. What is biosynthesis?

Ans. The formation of organic compound by living organisms is known as bisynthesis.

Q. 53. What do you mean by meristem culture?

Ans. Culture of apical meristems, particularly shoot apical meristem, for production of shoots and plantlets are known as meristem culture.

Q. 54. What is totepotency?

Ans. Potentiality or property of a cell to produce a whole organism is called as totepotency.

Q. 54. What is antisense therapy?

Ans. The in vivo treatment of a genetic disease by blocking translation of a protein with a DNA or an RNA sequence that is complementary to a specific mRNA is called antisense therapy.

19 PLANT TISSUE CULTURE

Q. 1. What is totipotency?

Ans. The cells or tissues are obtained from any part of the plant like stem, root, leaf etc. which are encouraged to produce more cells in culture and to express their totipotency i.e. their genetic ability to produce more plants.

Q. 2. What is the full form of HEPA?

Ans. High efficiency particulate air.

Q. 3. What is called the callus?

Ans. In culture, the excised plant tissues lose their structural integrity and change completely to a rapidly proliferation unorganised mass of cells which is called the callus.

Q. 4. What is explant?

Ans. The callus tissue, which is important to plant tissue culture, is produced experimentally from a small excised portion called the explant.

Q. 5. What is the full form of M.S. medium?

Ans. Murashige and Shoog medium.

Q. 6. What is organogenesis?

Ans. A whole plant can be regenerated in large number from callus tissue through manipulation of the nutrient and hormonal constituents in the culture medium. This phenomenon is known as plant regeneration or organogenesis or morphogenesis.

Q. 7. What is somatic embryogenesis?

Ans. By manipulation of nutrient and hormonal constituents, cluster of embryos can be achieved directly from the somatic cells of callus tissue. These embryos are called somatic embryos. This phenomenon is known as somatic embryogenesis.

Q. 8. What is caulogenesis?

Ans. Caulogenesis is a type of organogenesis by which only adventitious shoot bud initiation takes place in the callus tissue.

Q. 9. What is rhizogenesis?

Ans. Rhizogenesis is a type of organogenesis by which only adventitious root formation takes place in the callus tissue.

Q. 10. What is embryogenic potential?

Ans. The capability of the somatic plant cell of a culture to produce embryoids is known as embryogenic potential.

Q. 11. What is embryoid?

Ans. Embryoid is a small, well-organised structure comparable to the sexual embryo, which is produced in tissue culture of dividing embryogenic potential somatic cells.

Q. 12. What do you mean by artificial seeds?

Ans. Somatic embryo has no food reserves, but suitable nutrients could be packaged by coating or encapsulation to form artificial seeds, which behave like true seeds if grown in soil and can be used as a substitute of natural seeds.

Q. 13. What is single cell culture?

Ans. Single cell culture is a method of growing isolated single cell aseptically on a nutrient medium under controlled condition.

Q. 14. What is androgenesis?

Ans. Androgenesis is the *in vitro* development of haploid plants originating from totipotent pollen grains through a series of cell division and differentiation.

Q. 15. What is anther culture?

Ans. Anther culture is a technique by which the developing anthers at a precise and critical stage are excised aseptically from unopened flower buds and are cultured on a nutrient medium where the microspores within the cultured anther develop into callus tissue or embryoids that give rise to haploid plantlets either through organogenesis or embryogenesis.

Q. 16. What is pollen culture?

Ans. Pollen or microspore culture is an *in vitro* technique by which the pollen grains, preferably at the uninucleated stage, are squeezed out aseptically from the intact anther and then cultured on nutrient medium where the microspores, without producing male gametes, develop into haploid embryoids or callus tissue that give rise to haploid plantlets by embryogenesis or organogenesis.

Q. 17. What is protoplast culture?

Ans. Aseptic isolation and *in vitro* culture of protoplast of higher plant is called protoplast culture.

Q. 18. What is the full form of PEG?

Ans. Polyethylene glycol.

Q. 19. What is somaclones?

Ans. Plants derived from tissue culture has been variously referred to as somaclones or calliclones or protoclones.

Q. 20. What is micropropagation?

Ans. The regeneration of whole plant through tissue culture is called micropropagation.

20 PLANT BREEDING & BIOMETRY

Q. 1. Distinguish between autopolyploidy and allopolyploidy?

Ans.

Autopolyploidy	*Allopolyploidy*
1. It is the exact multiplication of the same genome.	1. Multiplication of single genome does not occur.
2. Same types of genome is multiplied, e.g. 6x, 4x etc.	2. Two distinct genomes are invalued i.e. the individuals are amphidiploids, e.g. (AB)

Q. 2. What is hybrid vigour?·Where do you find it?.

Ans. Unusual growth, strength and health of heterozygous hybrids from the less vigorous homozygous plants.

It is find in wheat.

Q. 3. What is inbreeding?

Ans. Mating between two individuals having a common ancestor is called inbreeding.

Q. 4. What is the difference between monosomic and nullisomic plants?

Ans. **Monosomic :** Monosomics are characterised by the absence of one chromosome i.e. each chromosome has its partner or homologous but one chromosome is without homolagous.

Nullisomic : These are a particular type of double monosomy where the two extra chromosomes are lacking the homologous chromosomes.

Q. 5. What is pure line selection?

Ans. A pure line consists of the progeny of a single self-fertilised homozygous plant and use for developing a variety. The method of production of a variety from the pure line is known as pure line selection.

Q. 6. What is acclimatisation?

Ans. The process of introduction of new plants into a locality or introducing the plants from their growing place to a new locality with different climate is termed plant introduction and their adjustment under this changed climate of the new locality is know as acclimatisation.

Q. 7. What is introgressive hybridisation?

Ans. In this type of hybridisation one species is completely replaced by another in nature if there is free inter-crossing among the plants of two species and population of both the species is equal at the beginning of hybridisation.

Q. 8. What is amphimixis?

Ans. (Gk *amphi* = both + *mixis* = mixing) It is the normal sexual reproduction in which the morphologically dissimilar male and female gametes unite together for the formation of zygate. Meiosis and fertilisation are the essential features.

Q. 9. What is clonal selection?

Ans. A method of selection of desirable clones from the mixed population of vegetatively propagated crop.

Q. 10. What is dockage?

Ans. Impurity percentage of a seed sample is called Dockage.

Q. 11. What do you mean by elite plants?

Ans. The plants which in trials and tests have proved genetically superior to other plants of the same species are called Elite plants.

Q. 12. What is emasculation?

Ans. In bisexual flowers, the removal of stamens before they burst and have shed their pollens. Its purpose is to check the self-pollination and is done before crossing.

Q. 13. What is exotic plant?

Ans. Foreign plant or variety i.e. the plant of foreign origin.

Q. 14. What is gamma garden?

Ans. The place where the plants or insect pests are treated with gamma rays emitted from the source is called gamma garden.

Q. 15. What is immunisation?

Ans. To induce the immunity in the host against a particular parasite.

Q. 16. What is inbred?

Ans. The production of pure lines in cross-pollinated plants is called inbreeds.

Q. 17. Distinguish between tetrasomy and nullisomy?

Ans.

Tetrasomy	*Nullisomy*
1. Add a single pair of chromosomes.	1. Lack of a single pair of chromosomes.
2. Chromosome formula 2n + 2	2. Chromosome formula 2n – 2
3. It adds homologous chromosome pairs.	3. It is lack of homologous chromosome pairs.

Q. 18. Cite the difference between pedigree selection and bulk selection.

Ans.

Pedigree selection	*Bulk selection*
1. This method is complicated and expensive	1. Simple, convenient and inexpensive.
2. Less time is required.	2. More time is required.
3. Here individual plants are selected and then bulked at end.	3. The selected plants are bulked at first and then individuals are tested separately.

Q. 19. What is ploidy breeding?

Ans. Many of our important crop plants like wheat, oat, sugarcane, cotton, tobacco, barley as well as many fruits and vegetables are cultivated by applying polyploids of various degrees. This is known as ploidy breeding.

Q. 20. What measurement does standard deviation give.

Ans. Standard deviation provides statistical measurement of biological error from calculated data.

Q. 21. What is genetic drift?

Ans. It refers to random fluctuation in gene frequenceis due to sampling error.

Q. 22. What is goodness of fit?

Ans. The degree to which the observed data and expected data are similar to each other. If the observed and predicted data are very similar, the Goodness of Fit is high.

Q. 23. Define standard error.

Ans. Standard error is a measure of variation of the mean from the standard deviation. It denoted $S\bar{n}$.

Q. 24. What is standard deviation?

Ans. Statistical measuring of the spread or variability of the sample around the mean value is known as standard deviation denoted as S.

Q. 25. What is degree of freedom?

Ans. In Chi-squre (X^2) test the number of classes to which an arbitrary number of observation can be assigned to yield the given total; ordinarily it is one less than the no. of classes in the data.

Q. 26. What is coefficient of interference?

Ans. One minus value of coefficient of coincidence.

Q. 27. What is chi-square test?

Ans. A statistical test for alternating whether observed data agree with an expected ratio

$$y^2 = \sum\left[(O-E)^2 / E\right]$$

Q. 28. What is genetic load?

Ans. The extent to which the burden of deleterious alleles causes a population to depart from its optimum fitness is called genetic load.

Q. 29. What is heterosis?

Ans. Heterosis refers to the superiority of F_1 hybrid in one or more characteristics over its parents.

Q. 30. What is laxurion?

Ans. Laxurion is the increased vigour and size of inter-specific hybrid vigour.

Q. 31. What are the theories of heterosis?

Ans. There are three main theories of heterosis. These are –

(i) Dominance.

(ii) Over-dominance.

(iii) Epistasis.

Q. 32. What is euploidy?

Ans. Euploidy means here variation of chromosome exactly multiple of the genome.

Q. 33. What is basic number?

Ans. Single set of chromosome is known as basic number or genome.

Q. 34. What is haploidy?

Ans. Half of somatic cell chromosome number is called haploidy. In higher organism the germ cell or reproductive cells chromosome number and in lower organism the body cell or gametophytic plants chromosome number is haploidy.

Q. 35. What do you mean by diploidy?

Ans. Here each and every chromosome is present in double, i.c. two in a set. But in higher organisms the body cell or somatic cell chromosome number and in lower organism the sporophytic cell chromosome number is called as diploidy.

Q. 36. What are the differences between autopolyploid and allopolyploid?

Ans.

Autopolyploid	*Allopolyploid*
1. Gigantism is found.	1. Hybrid vigour.
2. Plants are weak	2. Plant woody and strong.
3. Leaves are much larger.	3. Leaves are large but not like that of autopolyploid.
4. Number of fruit and flower less.	4. Number of fruit and flower more.
5. Susceptible to disease.	5. Resistant to disease.

Q. 37. What do you mean by autoallopolyploid?

Ans. Here autopolyploid and allopolyploid occur combinedly viz. both genomes are multiplied but one gene is less in number than the other.

Q. 38. What do you mean by non recurrent parent?

Ans. The parent to which hybrid material is crossed in a back cross is known as recurrent parent.

Q. 39. What is mass selection?

Ans. Mass selection refers to a method of crop improvement in which individual plants are selected on the basis of phenotype from a mixed population. Mass selection is applicable to both self- and cross-pollinated species.

Q. 40. What is artificial selection?

Ans. Artificial selection is to choose certain individual plants for the purpose of having better crop from a mixed population where the individual differ in characteristics.

Q. 41. What is the difference between monosomic and nullisomic?

Ans.

Monosomic	*Nullisomic*
1. Monosomic lack one complete chromosome.	1. Lack a single pair of chromosomes.
2. Chromosome formula 2n – 1.	2. Chromosome formula is 2n – 2.

Q. 42. What is plant introduction?

Ans. The process of introducing new plants into a locality or introducing the plants from their growing place to a new locality with different climate is termed plant introduction.

Q. 43. What do you mean by clonal selection?

Ans. A procedure of selecting superior clones from the mixed population of asexually propagated crop such as sugarcane, potato etc.

Q. 44. What are the differences between mass selection and pureline selection?

Ans.

Mass selection	*Pureline selection*
1. Used in both self- and cross-pollinated crop.	1. Used in self-pollinated species.
2. It is the simple, oldest method.	2. It is comparatively complex method.
3. Selection is based on the phenotype of plants.	3. Selection is based on the progeny of plant.
4. Mixture of several pureline in self-pollinated species.	4. It is progeny of a single homozygous.

Q. 45. What is trisomy?

Ans. Addition of a single chromosome to a set of homologous diploid chromosomes.

Q. 46. What is tetrasomy?

Ans. Addition of a pair of extra homologous chromosomes.

Q. 47. What is inbreeding depression?

Ans. Inbreeding depression is the loss or decrease in vigour and fitness as a result of inbreeding.

Q. 48. What are the differences between trisomy and tetrasomy?

Ans.

Trisomy	*Tetrosomy*
1. Here additlon of a single chromosome with a set of homologous chromosomes.	1. Here addition of a pair of extra homologous chromosomes.
2. The formula is 2n + 1	2. The formula is 2n + 2

Q. 49. What is standard deviation?

Ans. A statistical measuring of the spread or variability of the sample around the mean value is known as standard deviation :

$$\text{Standard deviation} \quad (SD)S = \frac{\sqrt{\sum fd^2}}{n-1}$$

Where f = frequency of individuals
(n – 1) = number of degrees of freedom
d = deviation of class value from mean

Q. 50. What is standard error?

Ans. Standard error is a measure of variation of the mean from the standard deviation.

$$\text{Standard error} \quad S\overline{n} = \frac{S}{\sqrt{n}}$$

Where S = Standard deviation
n = Total number of observation

Q. 51. What are the differences between mean and mode?

Ans. When frequency × class value is divided by whole population or number of population then it is considered as mean.

Mean is denoted as $\overline{x}$

$$\overline{x} = \sum \frac{fx}{n} \quad [f = \text{frequency, } x = \text{class value, } n = \text{population}]$$

The highest number of frequency of a class is mode.

Q. 52. What is hybridisation?

Ans. Hybridisation is the method of producing new crop varieties in which two or more plants unlike in genetical constitution are crossed.

Q. 53. What is the aim of hybridisation?

Ans. The aim of hybridisation is :

(i) To combine all the good characteristics into a single variety.

(ii) To increase the range of genetic variability by introducing various combination of character.

Q. 54. What do you mean by domestication?

Ans. The process of bringing wild and weedy species under human management.

Q. 55. What is epistasis?

Ans. Epistasis refers to the interaction between alleles of two or more different loci. Here is observed non-allelic interaction of three types viz. additive × additive, dominance × dominance, additive × dominance.

Q. 56. What is probability?

Ans. Probability may be define as the likelihood of the occurence of any particular form of an event and is regarded as the ratio of the number of way in which that form of event might occur to the total number of way in which that event might occur in any form.

Q. 57. What is hyperploidy?

Ans. Variation of the chromosome number due to addition of chromosome with the diploid number or multiple of basic number.

It is considered as (2n + z) z = any no. 1, 2, 3 etc.

Q. 58. What is hypoploidy?

Ans. Variation of the chromosome number due to reduction of chromosome number from the diploid or multiple of basic number.

It is considered as (2n – z) z = any no. 1, 2, 3 etc.

Q. 59. What are the different types of trisomics?

Ans. Trisomics are of three types :

(i) Primary trisomics : When the 3rd chromosome is like the other two chromosomes it is said to be primary trisomic.

(ii) Secondary trisomics : When the 3rd chromosome is not identical to the other two chromosomes it is called secondary trisomics.

(iii) Tertiary trisomics : When trisomic arose by the translocation between non-homologous chromosomes it is called tertiary trisomics.

Q. 60. What are the differences between autotriploid and autotetraploid?

Ans.

Autotriploid	*Autotetraploid*
1. They have three sets of chromosomes.	1. They have four sets of chromosomes.
2. In nature autotriploid are available as well as it may be formed by artificial method.	2. They are formed artifically.
3. Triploid are generally highly sterile. E.g. banana, surgarcane.	3. Tetraploids are generally fertile. E.g. Rye, coffe.

Q. 61. What are the two laws of probability?

Ans. (a) The probability of simultaneous occurence of two or more independent events is equal to the product of the probability that each will occur separately.

(b) The probability of separate occurences of two independent events is equal to the square root of the probability of their simultaneous occurence – provided that the two events are of equal probability.

Some exciting gifts are up for grabs with this book!!

SHORT QUESTION & ANSWER SERIES

BOTANY

Dibyendu Saha

FREE Gifts Rs. 1000/-

DEAR READERS

By now you must have gone through this book in the entirety. So we invite your assessment on this title with absolute impartiality and candid observations. After all, we publishers blindly depend on your benchmarking. Friends, please fill up the open spaces below with CAPITAL LETTERS. You may add some extra sheets of paper if the space provided is not sufficient for you.

i. What are the important topics of your syllabus which are not adequately covered in this book?

ii. In which chapters and on what topics the treatment of the subject-matter by the author is not satisfactory and why?

iii. Have you come across the typos, factual inaccuracies, inconsistencies, and other forms of mistakes in the book? If yes, then please specify the mistakes and page numbers.

iv. Name top three books on the same subject, which, in your consideration, are better than this book. Could you provide some reasons?

Your 1st Preference: ....

Your 2nd Preference: ....

Your 3rd Preference: ....

v. Any suggestion to improve the quality of illustrations, diagrams etc.?

vi. How did you like the cover design? Any suggestions on colours?

vii. Suggestions for further improvement of production standard:

viii. Who recommended you this book:

☐ Teacher ☐ Friend ☐ Parents ☐ Bookseller ☐ Librarian

ix. Name of the person who has recommended:

Designation: ....

Address: ....

Phone No.: ....

e-mail: ....

x. Name and address of the bookseller from whom you have purchased this book:

....

....

....

xi. Name and address of the library where you would like to find the book:

....

....

xii. Name and address of your institution (Please mention the University or Board):

....

....

xiii. Your name, e-mail, and complete postal address:

....

....

....

xiv. A few other titles published by us that you would like to read:

....

....

The best assessment will be rewarded half-yearly. The award will be in the form of our publications, as decided by the Editorial and Marketing Operations Division, amounting to Rs. 1000/- (Rupees one thousand) only.

....

Please mail the filled up coupon at your earliest to :

Editorial & Marketing Operations Division

NEW CENTRAL BOOK AGENCY (P) LTD

8/1 Chintamoni Das Lane, Kolkata 700 009

West Bengal, INDIA

Ph. No.: 033 + 22418411, 22418573, 22419094

Fax: 033 + 22419573, 22413852

e-mail: ncbapvtltd@eth.net, ncba@cal2.vsnl.net.in

Website: www.centralgroup.com